Handbook of
Thermoplastics Injection
Mould Design

For Chris and Cathy

Handbook of Thermoplastics Injection Mould Design

by

P.S. CRACKNELL and R.W. DYSON
London School of Polymer Technology
University of North London

Springer-Science+Business Media, B.V.

First edition 1993

© Springer Science+Business Media Dordrecht 1993
Originally published by Chapman & Hall in 1993
Softcover reprint of the hardcover 1st edition 1993

Typeset in 10/12 pt Times New Roman by Best-set Typesetter Ltd.,
Hong Kong

ISBN 978-94-015-7211-8   ISBN 978-94-015-7209-5 (eBook)
DOI 10.1007/978-94-015-7209-5

A catalogue record for this book is available from the British Library

Library of Congress Cataloging-in-Publication data available

∞ Printed on permanent acid-free text paper, manufactured in accordance
with the proposed ANSI/NISO Z 39.48-199X and ANSI Z 39.48-1984

Preface

Injection moulding is one of the most important methods of manufacturing plastics products. Through the development of sophisticated micro-processor control systems, the modern injection moulding machine is capable of producing precision mouldings with close tolerances in large numbers and with excellent reproducibility. This capability, however, is often limited by the lack of a proper appreciation of mould design.

The mould, or tool as it is often called, is at the heart of the injection moulding process. Its basic function is to accept the plastic melt from the injection unit and cool it to the desired shape prior to ejection. It is not, however, simply a matter of the mould having an impression of the shape to be moulded. Many other factors have to be taken into account – for example, the ability to fill the mould impression properly and efficiently without inducing weaknesses in the moulding and the efficient cooling of the moulding in order to maximise production rates without diminishing the quality of the moulding. In addition, the type of mould, gate and runner system, and ejection system which will best meet the needs of a particular job specification have to be determined. In our experience lack of attention to such factors leads to the mould limiting the ability of the injection moulding machine and preventing the process as a whole from achieving its true potential.

Injection moulds should not be designed in isolation but should be part of a team effort involving the product designer, the mould designer, the mould maker and the injection moulder. Each partner must be aware of the others' part in the process as a whole. This handbook will, it is hoped, be of some use to each of these partners.

P.S.C.
R.W.D.

Contents

1 Product and mould   **1**

  1.1 Introduction   1
  1.2 Shrinkage   1
      1.2.1 Mould shrinkage   1
      1.2.2 Sinking   3
      1.2.3 Internal stress and warping 4
  1.3 Wall thickness   5
  1.4 General features   7
      1.4.1 Taper and draft angle   7
      1.4.2 Corners   8
      1.4.3 Ribs   8
      1.4.4 Bosses   8
      1.4.5 Weld lines   9
  1.5 Undercuts   9
  1.6 Surface finish   10

2 Thermal considerations   **12**

  2.1 Introduction   12
  2.2 Heat considerations   12
  2.3 Specific enthalpy curves   14
  2.4 Cooling   15
  2.5 Cooling rates   16
  2.6 Mould cooling   19

3 Fluid flow   **21**

  3.1 Introduction   21
  3.2 Cooling systems   21
  3.3 Sample calculations   21
  3.4 Viscosity   22
  3.5 Limitations of Poiseuille's equation   24
      3.5.1 Turbulent flow   24
      3.5.2 Channel cross-section   25
  3.6 Polymer melts   27
  3.7 Examples of the use of flow equations and data   28
  3.8 Channel cross-section   30
  3.9 Filling the mould impression – other factors   31

**4 The injection moulding machine and its influence on mould
design**   **34**

  4.1 Introduction   34
  4.2 Machine function   34
      4.2.1 The machine base unit   34
      4.2.2 The injection unit   35
      4.2.3 The clamp unit   37
  4.3 Machine types and configurations   39
  4.4 Machine specification   41

5   Understanding moulds   **42**

    5.1   Introduction   42
    5.2   Tooling terminology   42
        5.2.1   The mould impression   42
        5.2.2   The mould core   42
        5.2.3   The mould cavity   43
        5.2.4   The split line   43
        5.2.5   Venting   45
        5.2.6   Ejection   46
        5.2.7   Back plate   47
        5.2.8   Sprue bush   47
        5.2.9   Register ring   48
        5.2.10   Tool location   48
    5.3   Mould types   48
    5.4   Choosing the correct mould   49
    5.5   Mould specification   51

6   The two-plate mould   **52**

    6.1   Introduction   52
    6.2   Tool construction   52

7   Runner and gate design   **56**

    7.1   Introduction   56
    7.2   Freeze flow characteristics   56
    7.3   Runner configurations   57
    7.4   Gate positioning and design   58
    7.5   Gate types   59
    7.6   Edge gate   59
    7.7   Fan gate   61
    7.8   Diaphragm gate   61
    7.9   Ring gate   62
    7.10   Spoke gate   63
    7.11   Tunnel or submarine gate   63
    7.12   Pin point gate   65
    7.13   Tab gate   65
    7.14   Flash or film gate   66

8   Mould cooling   **67**

    8.1   Introduction   67
    8.2   Cooling requirement   67
    8.3   Mould cooling time   69
        8.3.1   Estimation of cooling time   70
    8.4   Cooling media (or coolants)   71
    8.5   Conductive thermal properties of mould construction materials   71
    8.6   Cooling – design options   72
        8.6.1   Location of cooling channels   72
        8.6.2   Flat plane cooling   73
        8.6.3   Spiral cooling   73
        8.6.4   Finger or bubbler cooling   75
        8.6.5   Baffle cooling   75

9   Ejection   **77**

    9.1   Introduction   77
    9.2   Choice of ejection method   77

9.3   Ejection methods   79
      9.3.1   Pins and blades   79
      9.3.2   Ejector sleeves   80
      9.3.3   Valve headed ejectors   80
      9.3.4   Stripper ring and plate ejection   83
      9.3.5   Air ejection   83
9.4   Estimation of ejection force   84

10   The three-plate mould   86

10.1   Introduction   86
10.2   Why choose a three-plate mould?   86
10.3   Tool construction   87

11   Runnerless moulds   90

11.1   Introduction   90
11.2   Advantages of the hot-runner mould   90
11.3   Hot-runner systems   91
11.4   The externally heated hot manifold mould   91
11.5   The internally heated manifold mould   93
11.6   The insulated hot-runner mould   95

12   Undercut moulds   97

12.1   Introduction   97
12.2   Core pulling   97
12.3   Core pulling actuation methods   97
      12.3.1   Cam pin actuation   97
      12.3.2   Lost action cam pins   100
      12.3.3   Action wedges   101
      12.3.4   Hydraulic core pulling   102
      12.3.5   Pneumatic core pulling   104
      12.3.6   Electro-mechanical core pulling   105

13   Standard mould parts   107

13.1   Introduction   107
13.2   Why use standard mould parts?   107
13.3   Standard parts and assemblies   108
13.4   Standard elements   109
13.5   Accessory components   110
      13.5.1   Accessory cooling components   110
      13.5.2   Accessory ejection components   112
      13.5.3   Mould feeding systems   112
      13.5.4   Core pulling parts and accessories   113

14   Prototype moulds   114

14.1   Introduction   114
14.2   The case for a prototype mould   114
      14.2.1   Moulding aspects of the prototype component   114
      14.2.2   Mould design and constructional aspects highlighted by prototype
               moulding   115
14.3   Prototype mould tool materials and construction   117
14.4   Prototype tool construction   118
      14.4.1   Inserted bolster prototyping   118
      14.4.2   Modular prototype mould tool construction   120

15 Mould tool materials 121

15.1 Introduction 121
15.2 Mould construction material requirements 121
15.3 Mould construction materials 122
    15.3.1 Ferrous mould construction materials 122
    15.3.2 Alloy steels 122
    15.3.3 General purpose mould steels 123
15.4 Non-ferrous mould construction materials 125

Index 129

Abbreviations

Elements

Al	aluminium
Be	beryllium
C	carbon
Cr	chromium
Co	cobalt
Cu	copper
Mn	manganese
Mo	molybdenum
Ni	nickel
Pb	lead
Si	silicon
S	sulphur
V	vanadium
W	tungsten

Polymers

ABS	acrylonitrile–butadiene–styrene
CA	cellulose acetate
FEP	fluorinated ethylene propylene
HIPS	rubber modified polystyrene (=TPS)
LCP	liquid crystal polymer
PA 6	nylon 6
PA 6,6	nylon 6,6
PA 11	nylon 11
PA 12	nylon 12
PAE	polyarylate
PBT	polybutylene terephthalate
PC	polycarbonate
PE-HD	high density polyethylene (HDPE)
PE-LD	low density polyethylene (LDPE)
PES	polyethersulphone
PMMA	polymethyl methacrylate
POM	polyacetal (acetal)

PP   polypropylene
PPO   polyphenylene oxide (modified)
PPS   polyphenylene sulphide
PS   polystyrene
PSU   polysulphone
PVC   polyvinylchloride
PVC-P   plasticised PVC
PVC-U   unplasticised PVC (=UPVC)
SAN   styreneacrylonitrile
TPS   rubber modified polystyrene (=HIPS)

Miscellaneous

AISI   American Iron and Steel Institute
BS   British Standard
DIA   diameter
DIN   German standards
DRG   drawing
ISI   Swedish Standardisering Kommission
QA   quality assurance

1 Product and mould

1.1 Introduction

The mould impression is the part of the mould that accepts molten plastic from the injection unit either directly via a sprue or via a sprue and runner. It is in the impression that the molten plastic cools and assumes the desired shape. In designing an impression to make a particular product, a number of factors need to be considered. It is not simply a matter of making the impression conform precisely to the product design. Some factors, for example, are: the impression must be capable of being filled as easily as possible; the moulding must be as free of stress as possible; the plastic shrinks on cooling and the moulding must be removable from the open mould. While the mould designer may be able to allow for these factors in designing a mould impression to make a product, the designer must be capable of appreciating the limiting factors and, where necessary, must be able to advise the product designer of difficulties and suggest remedies. The following sections outline the salient factors which need to be considered.

1.2 Shrinkage

1.2.1 *Mould shrinkage*

All materials shrink on cooling due to thermal contraction. The shrinkage of the plastic on cooling from the melt temperature to the mould temperature is known, somewhat perversely, as the mould shrinkage. The principal reason for mould shrinkage is thermal contraction which is measured by the thermal expansion coefficient of the plastic. The expansion coefficients of plastics materials are high compared with metals (Table 1.1). Typically, a 100°C rise/drop in temperature will produce an increase/decrease of between 0.001 and 0.02 mm/mm depending on the material. Although this is small, it should not be ignored. Additionally, crystallisable thermoplastics shrink on crystallising, the amount of additional shrinkage depending upon the amount of crystallinity developed which in some polymers is very much dependent on the rate of cooling. For example, PET hardly crystallises at all when cooled rapidly unless it is seeded but slow cooling can produce up to about 50%

Table 1.1 Dimensional stability data

Material	Thermal expansion (mm/mm $K^{-1} \times 10^6$)	Mould shrinkage (%)	Water absorp. (%)
ABS (rigid)	80	0.3–0.8	0.3
Acetal	80	2.0–3.5	0.2
Cellulose acetate	100	0.3–0.7	2–6
Fluorinated ethylene propylene	90	3.0–6.0	0
Nylon 6,6	120	1.5–2.0	1.5
Nylon 6	100	1.0–1.5	1.6
Nylon 11	150	1.2	0.4
Nylon 12	104	1.0	0.3
Polybutyleneterephthalate	90	1.5–2.0	0.2
Polycarbonate	70	0.6–0.8	0.16
Polyethylene (LD)	170	2.0–3.5	0.02
Polyethylene (HD)	120	2.0–3.5	0.01
Polymethylmethacrylate	85	0.1–0.8	0.35
Polypropylene	110	1.5–2.5	0.01
Polyphenylene oxide (modified)	55	0.5–0.7	0.1
Polystyrene (GP)	70	0.2–0.6	0.2
Polystyrene (rubber modified)	120	0.2–0.8	0.2
Polyethersulphone	55	0.6–0.8	0.15
Polyvinyl chloride (rigid)	55	0.1–0.5	0.05
Styrene acrylonitrile	70	0.2–0.5	0.3
Steel	11–13		

The above are typical values for unfilled grades

crystallinity. The rate of cooling therefore determines the total amount of shrinkage as well as the properties of the product.

The use of fillers (mineral powders, glass fibres, etc.) can reduce the amount of shrinkage on moulding because they have much lower thermal expansion coefficients. However, processability may be adversely affected as well as dimensional stability.

It is common practice to quote a figure for *mould shrinkage* either in mm/mm or as a percentage for plastics materials (Table 1.1). Such figures should be regarded as indicative. The precise shrinkage observed will depend on *temperature drop, rate of cooling, shaping pressures* and *anisotropy due to orientation*.

Anisotropy arises primarily from molecular orientation produced during flow (chapter 3). The consequence is that shrinkage is greater in the flow (orientation) direction than in the cross-flow (transverse) direction. The difference in shrinkage depends on the material and the production methods. The inclusion of fibres also produces anisotropy. Since mineral fibres (glass and carbon) shrink less than plastics, this tends to negate the differential shrinkage of the plastic and at fibre loadings of above about 20%, it is common to find that the differential is reversed, i.e. shrinkage is greater in the transverse direction.

In dimensioning the mould, the mould dimensions should be slightly oversized compared with the product dimensions. The variability of shrinkage means that product tolerances should be as generous as other requirements permit otherwise tight control of the moulding process is required. Since amorphous materials shrink less than semi-crystalline materials, these materials are preferred where close product tolerances are necessary.

Other shrinkage factors which may need to be considered in dimensioning the mould are briefly listed below.

(a) The mould may not be at room temperature. Allowance should be made for the cooling of the moulding to room temperature. For amorphous plastics, this is simply a matter of thermal contraction and this may be estimated from the thermal expansion coefficient. Semi-crystalline materials may contract more due to further crystallisation on cooling.

(b) Even when cooled in the mould to room temperatures, semi-crystalline materials may shrink over a period of time after ejection as a result of further crystallisation. This is known as post-mould shrinkage which, though usually small (less than 0.01%), may have to be taken into account for high precision products. One remedy is to ensure that crystallisation is completed during moulding by, for example, increasing the cooling time.

(c) If the product is dimensioned for service at elevated temperatures, this may need to be taken into account when deciding upon mould dimensions.

1.2.2 Sinking

The moulding of thermoplastics in thick sections presents other shrinkage problems, especially where mould shrinkage is high (e.g. highly crystalline materials). In a thick section of an injection moulded product, for example, the outside layers in contact with the cold mould cool rapidly. The inside layers remain hot for longer because polymers are poor thermal conductors. As the centre layers cool, shrinkage occurs (often to a greater extent than the quickly cooled outer layers) and in doing so pulls the outer layers away from the mould wall causing sinking at the surface (Figure 1.1a).

In extreme cases (e.g. polypropylene) cavitation can occur as well as, or instead of, sinking. The inside of the cooling section shrinks away from itself producing voids or cavities which may result in essentially hollow centred sections which severely weakens the product (Figure 1.1b). Sinking is reduced by

• reducing part thickness;

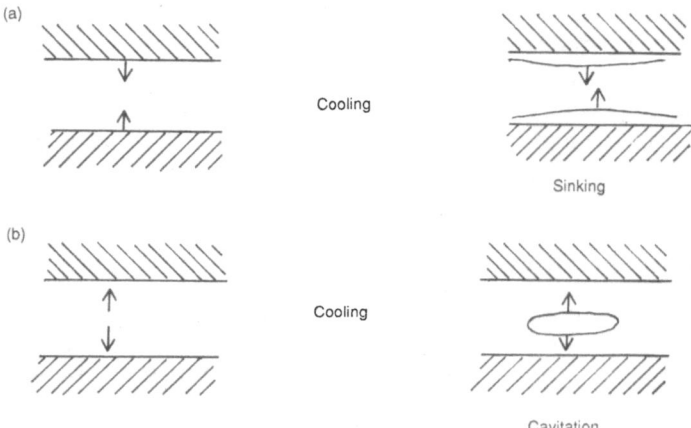

Figure 1.1 Showing forces producing sinking at the moulding surface and cavitation inside the moulding when thick sections are cooled from the melt temperature.

- incorporating fillers;
- maintaining an internal pressure during cooling.

The most convenient way of producing an internal pressure is by the incorporation of a gas in the melt. This can be done by dispersing a blowing agent (e.g. an azo-carbamide) into the polymer which decomposes during processing to provide the gas (nitrogen). The product is a slightly cellular moulding. An alternative is the gas (nitrogen) injection technique (e.g. Cinpress) which forms a tunnel of gas through the centre of thick sections. An advantage of the latter is that surface quality is not impaired through the escape of gas through the moulding surface as happens with blowing agents.

1.2.3 *Internal stress and warping*

When thick sections are moulded, rapid changes in temperature can cause thermally induced stresses due to differential expansion. If surface layers cool faster than the interior through poor conduction, the contraction of the surface will be greater than the interior thereby setting up stresses which can lead to warping and even failure in service at less than predicted stress levels. Thermal stresses induced during manufacture can be reduced by annealing during moulding or after moulding.

Internal stresses can also arise from flow induced anisotropy. Anisotropy arises from two principal sources: molecular orientation and the alignment of directional fillers such as fibres. Molecular orientation results from melt flow where the polymer chains are forced to change from their random coil state (ideally) to an elongated coil. The degree of elongation depends primarily upon the nature of the polymer and the shear rate (stress) experienced in flow. Shear rate depends on channel dimensions and increases as the channel cross-section decreases. Typical shear rates in practice are $1000-5000 \text{ s}^{-1}$ in injection moulding. The degree of coil distortion can be quite marked. On emerging from the channel (e.g. gate), the elongated coil will attempt to revert to the relaxed random coil state and if it can do so, the product will be isotropic. In practice, this ability to revert is hindered by

- loss of mobility in the polymer due to cooling;
- continued flow into the mould cavity.

The result is frozen molecular orientation in the general direction of flow. The degree of orientation is low compared with that which is deliberately induced in fibre and film production, but it can nevertheless be significant.

Frozen orientation produces stress which weakens the product and causes failure at lower applied stress levels since cracks can propagate more easily in the flow direction. However, stiffness is increased in the flow direction. Frozen orientation can also lead to dimensional instability. The application of heat induces molecular relaxation which produces warping or even gross distortion.

Molecular anisotropy can be minimised by using generous flow channels, low shear rates and slow cooling. Thin sections should be avoided. Fibres (usually glass) of length $0.3-0.5 \text{ mm}$ incorporated into thermoplastics as reinforcements increase the anisotropic effects described above because the fibres tend to orientate in the same flow direction as the polymer chains. Mineral powders of aspect ratio greater than unity (e.g. talc) also contribute to anisotropy but less so than fibres.

1.3 Wall thickness

Component wall thickness requires several competing factors to be taken into consideration. The mechanical requirements of a section may dictate a certain wall thickness. However, thick sections should be avoided because of the requirement to remove heat efficiently (the cooling part of the cycle is the time that limits the production rate), the need to avoid sinking, cavitation and warping. Thin sections should be avoided because of the need for the melt to flow easily to fill the cavity. Economic factors often dictate the final wall thickness selected (amount of material used, etc.). If a wall is too thin for the strength properties required, the wall can be reinforced with ribs (see below). Flow characteristics vary from one plastics material to another and Table 1.2 gives typical values for wall thicknesses.

Table 1.2 Wall thickness recommendations (mm)

Material	
ABS	1.00–3.50
Acetal	0.50–3.15
Acrylic	0.65–3.80
Cellulosics	0.65–10.00
Liquid crystal polymers	0.20–3.00
Long fibre plastics (Vertons)	1.90–25.00
Nylons	0.25–2.95
Polyarylate	1.15–3.80
Polycarbonates	1.00–3.80
Polybutylene terephthalate	0.65–3.20
Polyethylene (LD)	0.50–6.35
Polyethylene (HD)	0.75–5.00
Polyphenylene sulphide	0.50–4.55
Polypropylene	0.65–3.80
Polysulphones	1.00–3.80
Modified PPO	0.75–3.55
Polystyrene	0.85–3.80
SAN	0.85–3.80
UPVC	1.00–3.80

Table 1.3 Range of flow path ratios for some common plastics

Polymer	Flow path ratio
ABS	80–150
Acetals	100–250
Acrylic	100–150
Nylon 6	140–340
Nylon 6,6	180–350
Polybutylene terephthalate	160–200
Polycarbonate	30–70
Polyethylene (HD)	150–200
Polypropylene	150–350
Polystyrene	about 150
HIPS	about 130
Polysulphones	30–150
Polyvinylchloride (unplasticised)	about 60
Polyvinylchloride (plasticised)	up to 180
SAN	about 140

Courtesy of J.A. Brydson (1990) *Handbook for Plastics Processors*, Heinemann Newnes.

Wall thickness should be constant if possible. If not, the transition between thick and thin sections should be gradual because abrupt changes lead to internal stresses due to flow problems and different cooling rates. In general, the ratio of thick to thin should not be greater than 3:1. A further point is that in filling such a section, the flow must be from thick to thin.

For any section thickness, there will be a limit to the length of flow (flow path) possible. This is because as the plastic melt flows along a section, it cools producing an increase in melt viscosity and ultimately it sets up (hardens) thereby preventing the section from being completely filled. The ability to flow along a path depends on the path thickness, the type of plastics material and the particular grade (easy flow or stiff flow) of that material. It will also depend on moulding conditions. A guide to flow characteristics of materials in this respect is given by the *flow path ratio* which is the ratio of the length of flow possible per unit wall thickness. Table 1.3 gives the range of flow path ratios for some common plastics materials.

1.4 General features

1.4.1 *Taper and draft angle*

Components should be designed so that all surfaces perpendicular to the parting line are tapered so that the component can be ejected from the mould. The draft angle depends upon the material and the shape and depth of the moulding. The minimum draft angle is 0.5° with between 1.5° and 3° being normal. Thermoplastics shrink away from a wall cavity and onto the core on cooling. To ensure that the moulding remains in the correct part of the mould prior to ejection, correct selection of draft angles for interior and exterior surfaces is essential. For example, if the moulding is to remain on the core, the draft angle would be greater on the cavity side of the moulding. Draft angles for ribs and bosses should be at least 5°.

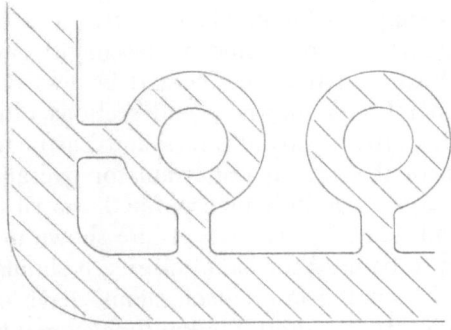

Figure 1.2 Section showing rounded corner and boss designs.

1.4.2 *Corners*

Corners should be rounded to avoid disruption of the melt flow as far as possible (Figure 1.2). Sharp corners disrupt the flow and this sets up internal stresses, especially with fibre reinforced materials. Furthermore, the polymer shrinks onto the inside of the corner. Stress concentration at corners can lead to failure of the component in service and flow induced stress encourages warping. The inside corner radius should preferably be at least half the wall thickness although a quarter of the wall thickness is acceptable. In any case, it should not be less than 0.5 mm. The outside corner radius is the internal radius plus the wall thickness unless other factors determine otherwise.

1.4.3 *Ribs*

Ribs are used as stiffening members along walls to avoid using thick sections to obtain the required stiffness. Ribs can cause problems such as stress concentration and sinking. Sink marks are prominent behind ribs where the effective section is thick. Ribs should therefore be thin, ideally between a third and a half of the wall thickness at the rib base. If a single thin rib does not give sufficient stiffening, a double rib can be used or, as an alternative, a thick rib can be used with a surface design feature to disguise the sink mark. The rib height is generally recommended to be between three and five times the wall thickness unless the rib is supported laterally. The number of ribs is also important and a number of small ribs is preferable to a single large rib to avoid distortion of the moulding on cooling. Ribs along a wall need not be identical with regard to height and width. The corner where the rib joins the wall should be rounded with a minimum radius of 0.2 mm.

1.4.4 *Bosses*

Bosses are used as supports for moulded inserts or as studs for use in assembling components. Bosses should be designed to avoid sinking and sharp corners. A boss wall thickness should be less than about three-quarters of the general wall thickness and sections other than circular should be avoided, partly to avoid stress and partly to reduce mould costs. The position of the boss is important for overall strength of the moulding and bosses are normally incorporated, via ribs if necessary, at corners or along wall sides. Typical designs are shown in Figure 1.2.

When the boss is to be used for metal inserts, it should be designed to receive the insert. The metal insert itself should have no sharp corners and where knurled inserts are used, the knurling should be rounded. In a properly designed system, the boss will shrink onto the insert on cooling because of the higher coefficient of expansion of plastics compared with metals. When the boss is to be used for self-tapping screws, particular attention needs to be paid to the inside diameter and wall thickness. A general recommendation is that the inside diameter should be the pitch of the screw and the outside diameter should be 2.5 times the screw diameter. The boss should be free-standing to reduce stresses when the screw is driven home.

From a mould design point of view, mouldings incorporating bosses should have ejector pins or sleeves incorporated at the base of each boss in the cavity to facilitate extraction. This also allows air to escape thus avoiding *burn* marks on the surface and incomplete filling of the cavity.

1.4.5 *Weld lines*

Weld lines occur where two flow fronts meet and always provide a source of weakness as the two fronts do not knit together. Weld lines often leave *witness marks* in the moulding surface. Two major sources of weld lines are multiple gating and flow around pins.

Multiple gating may be essential when large or complex mouldings are made. In such cases, welds are unavoidable but suitable gating allows the welds to be formed at points in the moulding where they can be best tolerated, e.g. points that are not likely to receive any major external stress. CAD systems help enormously in this respect.

Flow round pins used, for example, to create holes in the moulding provide weld lines downstream from the pin as the divided flow reunites. This region is therefore weakened. For this reason, holes are often machined in after moulding.

1.5 Undercuts

An undercut is a part of the mould where the molten plastic flows behind a retaining section thereby preventing the hardened plastic moulding from being easily ejected. Undercuts in general require moulds with moving parts to allow the product to be demoulded although parts may be stripped from the core provided that certain conditions are met. For rigid plastics, the part may be stripped if the undercut angle is not more than about 2°. Larger angles will induce stresses into the part on stripping even if the part is deformable enough to allow stripping at all. If the material is a softer plastic (e.g. plasticised PVC), then larger angle undercuts can be acceptable but even so, large undercuts will need hand stripping.

A screw thread is a common form of undercut. Products with screw threads usually require a mould with an unscrewing core in order that the

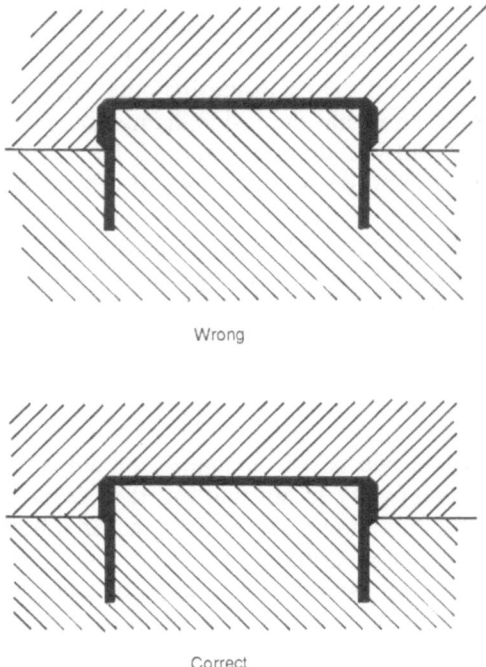

Figure 1.3 Mould section for jar cap showing parting line in incorrect and correct position to avoid undercut.

product can be demoulded without unscrewing the product from the mould by hand. Exceptionally, where the screw thread is of shallow depth and the material is sufficiently flexible, the part may be stripped from the core.

The problem of undercuts can often be solved by selecting the correct mould parting (split) line so that it avoids an undercut altogether. Figure 1.3 shows the essential features of a cylindrical cap for a jar being moulded with the parting line wrongly placed, producing an undercut, and correctly placed to avoid the undercut.

1.6 Surface finish

The finish required on the product determines the finish required on the mould surface. A high gloss finish to the product requires a mould of high quality nichrome steel or a tool steel with a highly polished chromed surface. Such moulds are expensive to produce because of the lengthy polishing treatments and the need still to maintain dimensional specifications. In some cases, textured surfaces are a design feature, either to achieve a particular effect or to mask problems resulting from moulding. Such problems arise, for example, when mouldings are made from expanded plastics or glass-filled plastics. Texturing a mould surface also increases the cost of the mould but the increase in cost can be minimised by using simple texturing.

2 Thermal considerations

2.1 Introduction

An injection moulding machine is a device for injecting a molten plastic material into a mould to shape it and solidify it. With thermoplastics, the material is fed to the barrel where heat is applied from an external source (usually electric heater bands) to produce a melt. Melt temperatures range from 150°C to 400°C depending on the material. The melt is subsequently injected into the mould which is at a temperature well below that of the polymer melt so that the melt cools and solidifies. Moulds are cooled by water circulating in channels in the mould block although sometimes other fluids such as oil are used. It is important to be able to calculate the amount of heat to be removed from the moulding so that proper cooling systems can be designed to give efficient moulding.

2.2 Heat considerations

A primary function of the barrel is to ensure that the plastic material is sufficiently fluid so that the melt is in the best condition for transfer to the mould. It does this by raising the temperature of the melt to an appropriate level for the particular material and the mould. Plastics materials each have their own processing requirements and in this respect each will have a characteristic melt temperature range. Precise temperatures used will depend on the grade of polymer on which the plastic is based, the type and quantity of any additive present and the mould filling requirements. In the following sections, quoted temperatures should be regarded as typical. Tables 2.1 and 2.2 show typical temperatures for some common amorphous and semi-crystalline materials.

Consider first the amorphous polymers. The amount of heat required to get them into a suitable molten state depends on the melt temperature and on the specific heat of the polymer. The specific heat is the amount of heat required to raise the temperature of one kilogram of material by 1°C. Thus the amount of heat per kilogram to raise the temperature from ambient (say 20°C) is

$$\text{heat (kJ)} = \text{specific heat (kJ/kg)} \times (\text{melt temperature} - \text{ambient}) \quad (2.1)$$

Table 2.1 Amorphous polymers

Polymer	Melt temp. (°C)	Specific heat (kJ/kg/°C)	Heat required (kJ/kg)
PVC (rigid)	180–200	1.05	168
Polystyrene (GP)	210–250	1.34	268
HIPS	210–250	1.40	280
ABS	210–260	1.40	308
Noryl	240–300	1.34	308
Polycarbonate	280–320	1.26	353

Table 2.2 Semi-crystalline polymers

Polymer	Melt temp. (°C)	Specific heat (kJ/kg/°C)	Latent heat (kJ/kg)	Crystallinty (%)	Heat required (kJ/kg)
LDPE	190–220	2.30	150	50	489
HDPE	210–240	2.30	209	80	627
Acetal	180–215	1.45	163	80	420
PP	210–250	1.93	100	75	499
PA 6	240–280	1.59	130	50	423
PA 66	275–285	1.67	130	50	491

Table 2.1 gives specific heat values and heat requirements for the polymers listed. It should be noted that PVC has both a lower melt temperature requirement *and* a much lower specific heat than the others thereby producing the lowest heat requirement by far.

Consider now the semi-crystalline polymers. There is an additional heat requirement to that of simply calculating the amount of heat required to raise the temperature from ambient to the processing temperature. This is the heat required to melt the polymer crystals at their melting point and this heat is called the *latent heat of melting*. It is defined as the amount of heat required to melt 1 kg of crystalline polymer at its melting point. Values are listed in Table 2.2. Crystalline polymers are not 100% crystalline and the amount of crystallinity can vary between 30% and 80% depending on the polymer and its thermal history. Typical percentage crystallinities for granular moulding materials are given in Table 2.2. The total heat requirement is therefore the amount of heat to raise the temperature according to the specific heat equation above *plus* the latent heat requirement which is given by

latent heat requirement (kJ) = latent heat (kJ/kg) × % crystallinity

(2.2)

A comparison of the heat requirements of the polymers in Tables 2.1 and 2.2 shows that semi-crystalline polymers generally have higher heat requirements than wholly amorphous polymers and this is because of the extra heat needed to melt the crystals.

The heat values given in Tables 2.1 and 2.2 are too low because, in the interests of simplicity, several factors have been ignored. These include the fact that specific heat is not a constant but varies with temperature and the amount of crystallinity and the fact that the polymer melt specific heat will be different from that of solid semi-crystalline polymers.

2.3   Specific enthalpy curves

In practice, heat requirements may be better estimated using specific enthalpy curves (Figure 2.1). Enthalpy is another name for heat. These curves plot the amount of heat required to raise the temperature of one kilogram of polymer from 20°C to any temperature up to the normal processing range. Two distinct types of curve can be seen. One type is gently curving but almost linear and belongs to amorphous polymer. The other type shows a marked steepening followed by a sharp transition to a linear part at the highest temperatures. These curves belong to the semi-crystalline polymer and the melting range and melting point can be determined from the curve.

Melting begins where the curve begins to steepen and the melting point is the temperature where the sharp transition occurs. If the polypropylene

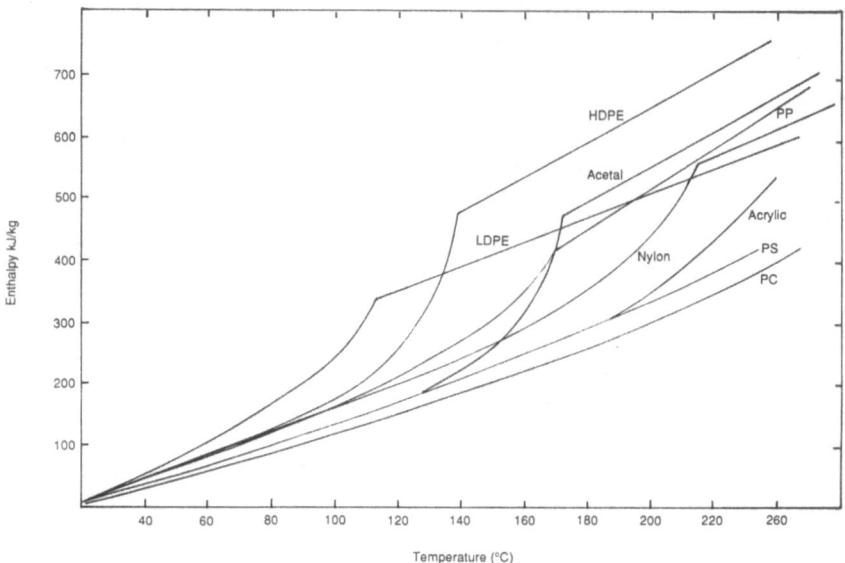

Figure 2.1 Enthalpy curves for some common polymers.

curve is considered, melting begins at about 120°C and the melting point is at about 170°C.

The curves are used as in the following example to determine the total heat requirement to raise the temperature of a polymer to its processing temperature. Consider polystyrene. The amount of heat required to raise 1 kg of polymer to 220°C is found by reading off from the enthalpy axis the enthalpy value at the point where the polystyrene line cuts the 200°C line, i.e. 380 kJ. Similarly, polypropylene at 240°C requires 670 kJ/kg.

A point to note is that the polymer melt is normally at temperatures up to 20°C above the heater control temperature settings as a result of shear heating produced by the screw rotation.

Finally, if enthalpy curves are not available for a particular polymer or grade of material, heating requirements can be estimated using equations (2.1) and (2.2).

2.4 Cooling

When the hot melt reaches the mould cavity, the heat has to be removed from the moulding. In many cases, the mould will be at or about ambient temperatures (say 20°C). The amount of heat to be removed per moulding or shot can be calculated using the enthalpy curves as before. For example, let us suppose that the total shot weight of polypropylene melt at 240°C is 100 g and it is required to be cooled to 40°C in a cold mould before ejection. The heat required to be removed is the difference in the enthalpy values obtained from the polypropylene curve at 240°C and 40°C. These values are 670 kJ/kg and 35 kJ/kg respectively. The difference is 635 kJ/kg and therefore the amount of heat per shot is $100/1000 \times 635$ kJ = 63.5 kJ per shot. It might be that the mould is at 80°C to allow for some annealing. In this case, the amount of heat required to be removed is $670 - 115 = 555$ kJ/kg which means that 55.5 kJ need to be removed per shot. Using a hot mould therefore requires less heat to be removed. Even so, the amount of heat to be removed is considerable and it is this fact which makes the cooling part of the mould cycle the longest part of the cycle in most cases. In order to keep this part of the cycle as short as possible, cooling should be as efficient as possible. Since polymers have low thermal conductivities, moulding sections should be as thin as possible to allow the heat to be conducted away as rapidly as possible. Section thickness of the moulding is also determined by stiffness requirements of the product and the need to fill the section with melt.

Table 2.3 gives a more complete list of thermal data for polymeric and non-polymeric materials. It is apparent that compared with metals, glass and fillers, polymers have high specific heats. It is common to incorporate

Table 2.3 Thermal properties of some polymers (typical values only)

Polymer	Specific heat (kJ/kg.K)	Latent heat (kJ/kg)	Conductivity (W/m.K)	Density (g/cm³)
ABS (rigid)	1.40	–	0.12	1.07
Acetal	1.45	163	0.23	1.42
Cellulose acetate	1.51	–	0.25	1.28
Nylon 6,6	1.67	130	0.25	1.14
Nylon 6	1.68		0.25	1.12
Nylon 11	2.40		0.42	1.04
Nylon 12	2.10		0.29	1.02
Polycarbonate	1.26	–	0.19	1.21
Polyethylene (HD)	2.30	209	0.49	0.96
Polyethylene (LD)	2.30	150	0.34	0.92
Acrylic	1.47	–	0.20	1.18
Polypropylene	1.93	100	0.14	0.91
PPO–mod	1.34	–	0.26	1.06
Polystyrene	1.34	–	0.12	1.06
HIPS	1.4	–	0.13	1.05
PVC rigid	1.00	–	0.25	1.41
SAN	1.4	–	0.12	1.08
Glass	0.67		0.9	
Mild steel	0.42		63	
Aluminium	0.91		200	
Brass (70:30)	0.37		385	
Copper	0.38		410	
Beryllium			230	

Note: compounds containing fillers will have values significantly different from the above values.

glass fibres and powder fillers into thermoplastics to improve strength, stiffness and dimensional stability. A consequence of this is that such compounded plastics will have lower heat capacities than the parent polymers. Consequently, although melt temperatures may be a little higher, the heat requirements of such compounds will be lower and, more importantly, less heat will have to be removed.

Consider a 30% glass-filled polypropylene. The heat required to raise the temperature from 20°C to 240°C is 513 kJ/kg which is considerably less than that for polypropylene alone.

It should also be appreciated that glass fibres and powder fillers such as chalk and talc have higher thermal conductivities than polymers. Therefore filled compounds will gain or lose heat more rapidly than unfilled polymer.

2.5 Cooling rates

The cooling rate depends upon the rate of heat flow from the moulding to the mould. Since the heat flow is by conduction through the solidifying

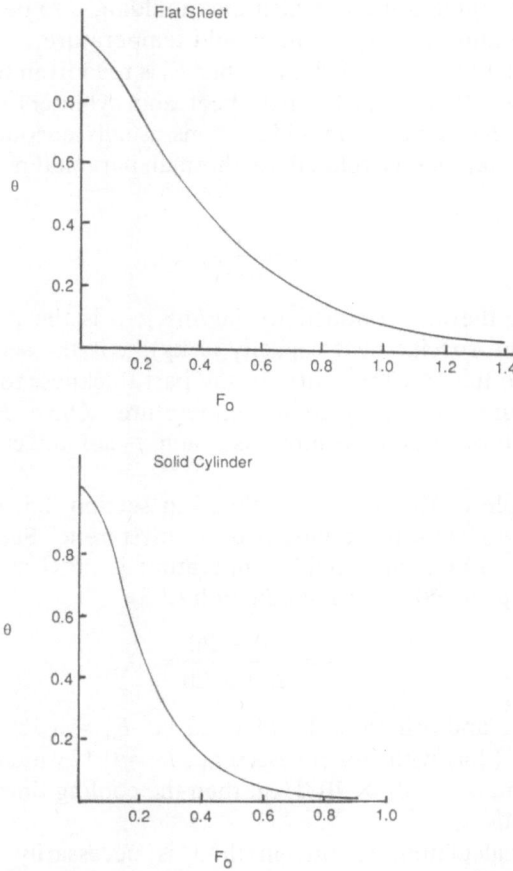

Figure 2.2 Dimensionless temperature factor as a function of Fourier number.

polymer, the cooling rate will depend upon the thermal conductivity (K) as well as the melt temperature, mould temperature and moulding thickness. Since the moulding temperature is decreasing throughout cooling, the simple steady state heat flow equation for conduction is inapplicable. However, the use of thermal parameters in conjunction with the Fourier number curves (Figure 2.2) enables a reasonable estimate of the cooling time to be obtained without too much difficulty. The method is as follows:

The dimensionless temperature (θ) is first calculated from equation (2.3)

$$\theta = \frac{T_E - T_m}{T_M - T_m} \tag{2.3}$$

where T_E is the temperature at which the moulding is to be ejected; T_M is the melt temperature and T_m is the mould temperature.

Having calculated θ, the Fourier number F_0 is read from the appropriate curve in Figure 2.2. Curves for flat sheet and cylinders only are given since these are the shapes of mould sections usually encountered.

The Fourier number is related to thermal parameters and thickness dimension by

$$F_0 = \frac{4Kt}{\rho \sigma x^2} \tag{2.4}$$

where K is the thermal conductivity (W/mK); ρ is the polymer density (kg/m³); σ is the specific heat capacity (J/kg); x is the section thickness (m); and t is the time for the centre of the part thickness to cool from the melt temperature to the ejection temperature. Once F_0 is obtained for a given set of cooling conditions, then t can be calculated using equation (2.4).

As an example of the method outlined in section 2.5, consider a flat plate of thickness 2 mm to be moulded in polystyrene. Suppose the melt temperature is 220°C, the mould temperature is 20°C and the ejection temperature is to be 60°C. Using equation (2.3),

$$\theta = \frac{60 - 20}{220 - 20} = 0.2$$

From Figure 2.2 and using the flat plate curve, $F_0 = 0.72$.
Using equation (2.4) with, for polystyrene, $K = 0.12$ watts/mK; $\rho = 1.06 \times 10^3$ kg/m³; and $\sigma = 1.34 \times 10^3$ J/kg, then the cooling time t is calculated to be 8.5 seconds.

Any value calculated by this method is necessarily an estimation because K, σ and ρ vary with temperature. Other assumptions are made in the theory behind this approach, principally that contact between the melt and mould wall is perfect which is only approximately true at best. Nevertheless, it is simple and gives a reasonable estimate. A practical problem is that melt temperatures are often not precisely known and are often above the set temperatures of the barrel. If it is assumed that the real melt temperature of the polystyrene in the above calculation is 20°C above the set temperature of 220°C, then t is calculated to be 9.5 seconds.

Table 2.4 shows some calculated cooling times for various melt, mould and ejection temperatures for the above polystyrene moulding. This illustrates quantitatively how changing the temperatures affects the cooling time. It will be clear that the colder the mould, the more rapid the cooling thereby leading to faster cycle times. However, rapid cooling can present problems of dimensional stability and internal stress in the moulding. If these are unacceptable, the mould temperature may need to be high (60°C and above) to anneal the product and it may be this

Table 2.4 Calculated cooling times

Melt temp. (°C)	Mould temp. (°C)	Ejection temp. (°C)	Cooling time (s)
220	20	60	8.5
240	20	60	9.5
220	20	30	16.5
220	10	60	4.0

requirement that determines the 'cooling time'. The residence time in the mould and the mould temperature will depend on the polymer used, part thicknesses and the degree of stress relief required. Before embarking on any calculations, therefore, it is wise to ascertain and not assume the mould temperature to be used in production.

In practice, a particular moulding may vary in wall thickness. Where this is so, the thickest section should be used when estimating cooling time.

2.6 Mould cooling

In order to cool the moulding, the heat transferred from the moulding to the mould needs to be removed from the mould. This may be done by passing water through channels suitably positioned in the mould. If the amount of heat required to be removed per shot and the cooling time are known, then the heat removal rate is obtained and the coolant flow rate required to maintain the mould temperature can be calculated. In doing this, it is necessary to specify a permissible temperature variation in the coolant as it flows through the mould. It is usual to specify $\pm\,2$°C; that is, the coolant enters the mould at 2°C below the nominal mould temperature and leaves at 2°C above it.

Consider the polystyrene plate moulded under conditions described in Table 2.4, first temperature settings and requirements. The calculated cooling time is 8.5 s. Suppose the plate is 10 cm square. The volume of the plate is

$$(10 \times 10^{-2})^2 \times (2 \times 10^{-3}) \text{ m}^3 = 2 \times 10^{-5} \text{ m}^3.$$

The mass is therefore

$$(2 \times 10^{-5}) \times (1.06 \times 10^3) \text{ kg} = 0.0212 \text{ kg}$$

Using the specific enthalpy curves (Figure 2.1), the amount of heat required to be removed in cooling polystyrene from 220°C to 60°C is

$$370 - 60 = 310 \text{ kJ/kg}.$$

The amount of heat to be removed per moulding is therefore

$$310 \times 0.0212 = 6.6\,\text{kJ} = 0.773\,\text{kJ/s}$$

The mass of water needed to remove the heat at this rate is found by equation (2.1) using the temperature limits of $\pm$ 2°C and given that the specific heat of water is 4.18 J/g (4.18 kJ/kg), viz:

$$0.773 = 4.18 \times \text{mass of water} \times 4$$

The mass of water per second is therefore

$$0.047\,\text{kg/s} = 4.7 \times 10^{-5}\,\text{m}^3/\text{s}$$

which is approximately 0.05 l/s or 3 l/min.

Such a calculation allows the dimensions of the cooling channels to be calculated (see chapter 3).

3 Fluid flow

3.1 Introduction

There are two aspects of fluid flow that are of significance in mould design. One is the flow of cooling liquids through the mould to remove the heat from the moulding prior to ejection, the other is the flow of the polymer melt through the runner system to fill the cavity. Since the flow of simple liquids such as water is simpler than that of polymer melts, their flow will be considered first.

3.2 Cooling systems

The cooling medium, usually water but sometimes oil, is pumped from the chiller/heater unit through a flexible delivery hose to the mould assembly where it passes through channels drilled in the metal before being returned to the pump unit or dumped to waste via another flexible hose. The two halves of the mould cavity are connected by flexible hoses. The rate of removal of the heat from the moulding depends on the coolant flow rate through the cooling channels. The cooling channels are normally drilled in the metal and are therefore circular in cross-section. The coolant flow rate depends on the flow channel dimensions, i.e. the length (L) and the radius (r), the viscosity (η) of the coolant and the pressure (P) causing flow. These factors are related in Poiseuille's equation for flow through circular channels:

$$Q = \frac{P\pi r^4}{8\eta L} \tag{3.1}$$

with units in metres, kilograms and seconds.

Equation (3.1) enables calculations to be done to ensure that the coolant channels are properly dimensioned for a given pumping system or that the pumping system is adequate for a particular cooling channel system.

3.3 Sample calculations

Suppose the polystyrene moulding (section 2.6) is to be cooled with a flow rat of 31/min. What is the maximum length of flow channel available using a pressure of 10 psi (70 000 Pa) if the channel diameter is 6 mm, which is a typical bore? Rearranging equation (3.1), we get

$$L = \frac{P \pi r^4}{8 \eta Q} \tag{3.2}$$

and substituting the data bearing in mind that viscosity of water is 0.001 Pa s; 3 l/min = 0.00005 m³/s; diameter 6 mm = radius 3 mm = radius 3×10^{-3} m

$$L = \frac{70\,000 \times \pi \times (3 \times 10^{-3})^4}{8 \times 0.001 \times 5 \times 10^{-5}} = 44\,\text{m}$$

which is more than ample for such a mould.

Alternatively, suppose that the pump is operating at 20 psi (140 000 Pa), what is the smallest channel diameter that can deliver 3 l/min over a 10 m length? Rearranging equation (3.1) again

$$r^4 = \frac{8 \eta Q L}{\pi P} \tag{3.3}$$

and substituting the data

$$r^4 = \frac{8 \times 0.001 \times 5 \times 10^{-5} \times 10}{\pi \times 140\,000} = 9.09 \times 10^{-12}$$

$r = 1.7 \times 10^{-3}$ m = 1.7 mm and the diameter is 3.4 mm.

Poiseuille's equation is a simple and effective way of calculating flow channel parameters and any one parameter from channel diameter, channel length, pressure drop and flow rate can be calculated provided that the other three are specified and the fluid viscosity is known.

It is common to use delivery hose of larger bore than the cooling channels as the larger bore produces a smaller pressure drop across the hose thereby making more pressure available to produce a given flow rate across the cooling channel of the mould. This means that a greater length of mould cooling channel can be utilised if necessary. Typically, the delivery hose for a 6 mm channel would be 12.7 mm (0.5 inches) in diameter.

Calculations of cooling channel dimensions cannot give information as to where the cooling channels should be placed in the mould. This is largely a matter of experience and common sense and is dealt with in chapter 8.

3.4   Viscosity

With regard to fluid viscosity, a value of 0.001 Pa s has been used for water in the above calculations. This value is correct for water at 20°C.

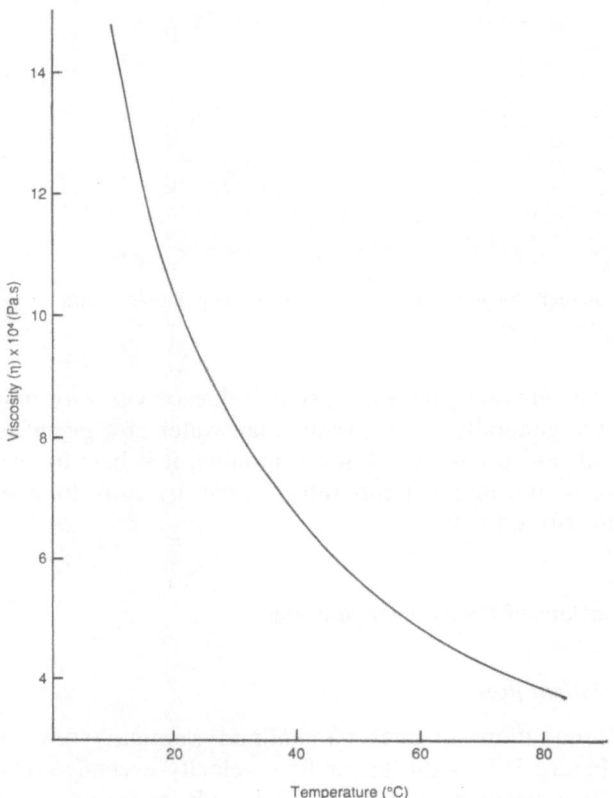

Figure 3.1 Variation of the viscosity of water with temperature.

Viscosity decreases exponentially with increase in temperature and Figure 3.1 shows the variation of viscosity with temperature for water. The general equation for temperature variation of viscosity is

$$\eta = A \exp(E/RT) \qquad (3.4)$$

where E is a constant characteristic of the liquid, $R = 8.314$ J/mole K and T is the temperature in absolute terms (K). It is often more useful to plot log η as a function of $1/T$ as this produces a straight line which allows interpolation and extrapolation of data.

If the mould cooling water is above room temperature, the correct viscosity for calculations can be obtained from the curve. However, if calculations are done using the value at 20°C, then channel sizes so calculated will be more than adequate since the lower viscosity at higher temperatures will tolerate smaller diameters or longer channels for the same flow conditions (P and Q). If the cooling medium is some other

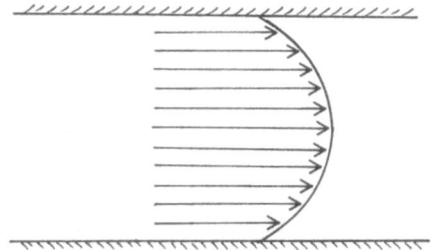

Figure 3.2 Laminar flow in a circular channel showing parabolic nature of velocity profile.

fluid, e.g. an oil, then the appropriate value of viscosity must be used. Since oils are generally more viscous than water at a given temperature, and since oils are usually used in hot moulds, it is best to use a viscosity appropriate to the mould temperature range by consulting the viscosity temperature curve for the oil.

3.5   Limitations of Poiseuille's equation

3.5.1   *Turbulent flow*

Poiseuille's equation assumes streamlined (laminar) flow of the type shown in Figure 3.2. If the linear flow velocity exceeds a certain value, then the flow becomes turbulent. Reynolds equation (3.5) gives the critical velocity (v) for turbulence in terms of the tube radius (r), the liquid density (ρ) and viscosity (η) and a constant known as the *Reynolds number* (R) which may be taken as 2000.

$$v = \frac{R\eta}{\rho r} = \frac{2000\eta}{\rho r} \tag{3.5}$$

Substitution of the data for water at 23°C flowing through a 6 mm channel will show that the flow is turbulent. In fact, water flow through mould cooling channels is usually turbulent and although Poiseuille's equation does not strictly apply, its simplicity allows good estimates of channel sizes to be made. Turbulent flow is more efficient at cooling than streamlined flow and is therefore desirable. The much higher viscosities of oils used as the cooling medium means that the critical flow velocity is often not exceeded unless high temperatures are used (reducing the viscosity) and oil flow is generally streamlined.

3.5.2 *Channel cross-section*

Poiseuille's equation is useful when considering the flow of simple liquids such as water and oils in circular channels. It is clearly of no use in channels of non-circular section or for liquids such as polymer melts whose flow behaviour is not simple. A more general equation is needed.

It is assumed that with simple liquids, flow is streamlined and flow is brought about by shear forces (pressure) causing the liquid to flow in layers. The layers nearest to the channel walls move slowly through frictional drag of the stationary wall and the layers in the centre of the channel move with the greatest velocity. The liquid is sheared in flow with a velocity profile which is parabolic (Figure 3.2). The parameters which characterise the flow are shear stress (τ) and shear rate ($\dot{\gamma}$) which are related in the general equation describing flow

$$\tau = \eta\dot{\gamma} \tag{3.6}$$

where η is the fluid viscosity or resistance to shear.

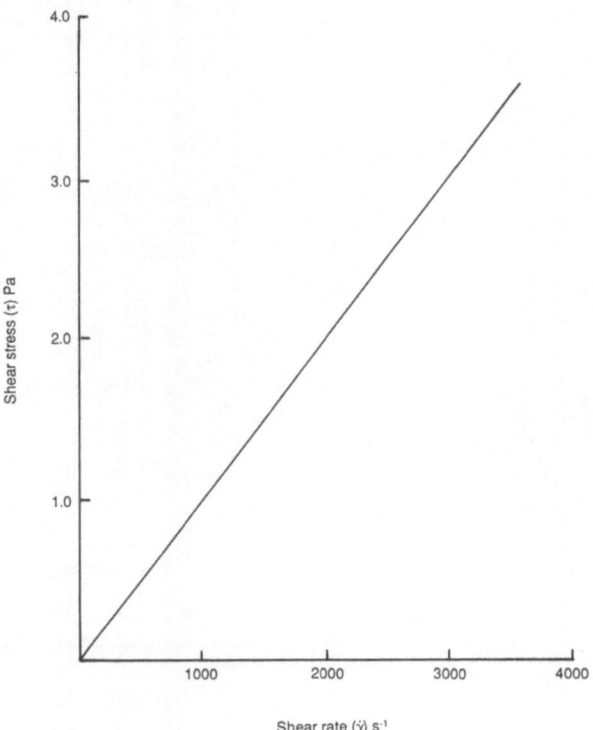

Figure 3.3 Newtonian flow behaviour. Data for water at 20°C.

For a circular cross-section channel,

$$\tau = \frac{Pr}{2L} \quad \text{and} \quad \dot{\gamma} = \frac{4Q}{\pi r^4} \tag{3.7}$$

where the symbols have the meanings already specified. If equations (3.5) and (3.6) are combined and rearranged, then Poiseuille's equation can be obtained.

The implication of equation (3.3) is that for a liquid of specified viscosity, the shear rate is directly proportional to the shear stress and a straight line results when shear stress is plotted as a function of shear rate (Figure 3.3). Such behaviour is said to be Newtonian and such fluids are known as Newtonian fluids.

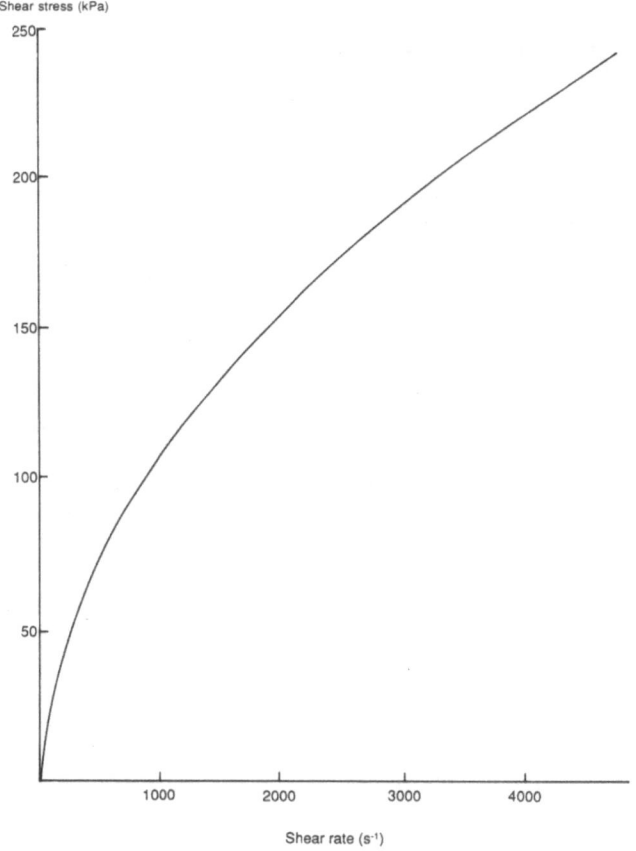

Figure 3.4 Showing the pseudoplastic behaviour of a polymer melt. Data for a polypropylene at 210°C.

3.6   Polymer melts

Polymer melts are not Newtonian in behaviour but exhibit shear thinning characteristics (also known as pseudoplastic behaviour) when subjected to increasing shear forces. In simple terms, this means that the flow becomes relatively easier as the pressure is increased, that is the viscosity reduces as the flow rate increases. This behaviour is shown for a typical polymer melt, polypropylene, in Figures 3.4 and 3.5.

Equation (3.5) is modified to describe this behaviour and becomes

$$\tau = K\dot{\gamma}^n \qquad\qquad (3.8)$$

where K is the consistency factor and n is the power exponent with values less than unity for shear thinning liquids.

The reason for shear thinning is the long chain structure of the

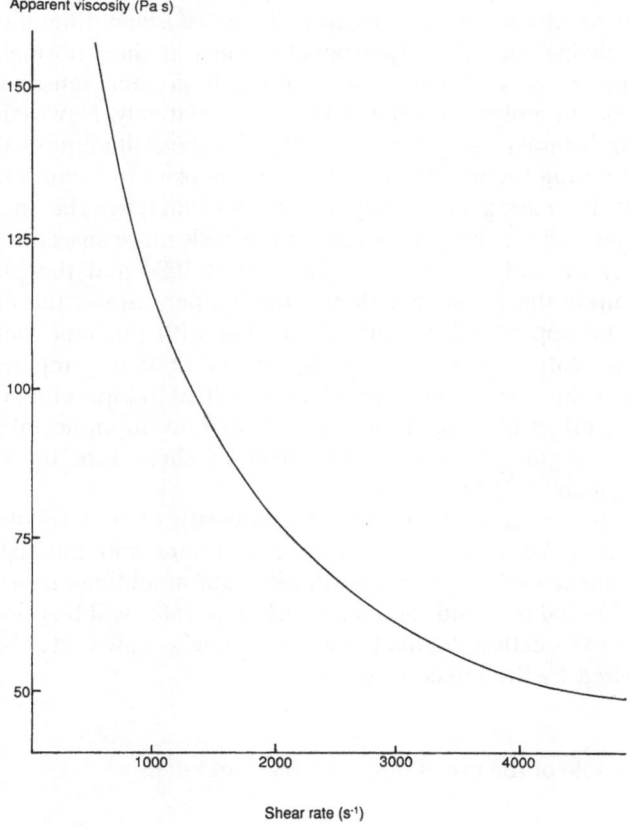

Figure 3.5 Showing the dependence of viscosity on shear rate – a polypropylene at 210°C.

Table 3.1 Typical viscosity date for an acetal copolymer

Temperature (°C)	195	205	220	
Apparent viscosity (Pa s)	210	190	163	
Shear rate constant at $1000\,N\,s\,m^{-1}$				
Shear rate (s^{-1})	10^2	10^3	10^4	10^5
Apparent viscosity (Pa s)	440	190	81	63
Temperature constant at 205°C				

polymers. The shearing of the melt during flow causes the polymer chains, which at rest are essentially random coils, to deform and elongate to more streamlined shapes. Since the streamlined shapes can slide past each other more easily, melt flow is easier and the viscosity of the melt reduces. Increasing the shear rate causes more streamlining and so increasing the shear rate reduces the viscosity.

The degree of shear thinning, which is indicated by n in equation (3.7), depends upon the polymer. Inherently flexible polymer chains such as polyethylene chains show a greater degree of shear thinning than rigid polymer chains such as polycarbonate. Thus at their normal processing temperatures, polyethylenes show a much greater sensitivity towards shear than do polycarbonates which are relatively Newtonian in their behaviour. Temperature also has an effect on shear thinning characteristics. Since increasing the temperature reduces viscosity by increasing the chain flexibility, increasing melt temperatures will increase the shear thinning effect making the volume flow rate of the melt more susceptible to shear. In general for polymer melts, $0.25 < n < 0.80$ and the precise value depends upon the polymer melt and the temperature of the melt.

It will be apparent from the above that with polymer melts it is not possible to state a value for viscosity on the basis of temperature alone. Viscosity values must be quoted at specified temperatures and shear rates. The effect of temperature and shear rate on an acetal is shown in Table 3.1. Figure 3.5 shows the effect of shear rate on viscosity for polypropylene at 210°C.

It will be apparent that the melt viscosity of a particular grade of polymer is a function of the melt temperature and the shear rate. In practice, shear rates experienced in injection moulding are usually in the range $1000-5000\,s^{-1}$ and the maximum shear rates will be observed in the smallest cross-section channels such as runners, gates, etc. Shear rate is also affected by the injection speed.

3.7 Example of the use of flow equations and data

An important factor in mould design is the correct dimensioning of runners. It is accepted practice that the pressure drop across a runner

Table 3.2 Some typical viscosity data for general purpose or standard grades of polymers at $1000\,s^{-1}$ shear rate

Polymer	Temperature (°C)	Viscosity (Pa s)	Temperature (°C)	Viscosity (Pa s)
PSU	360	643	380	516
PES	360	320	380	230
PPS (40% GF)	320	185	360	115
PPO (mod)	280	145	320	90
PPO (20% GF)	280	229	320	142
PET (20% GF)	270	275	295	160
PEEK	360	290	–	–
PC	300	420	360	100
PBT	240	317	280	43
PA6	280	89	300	69
ABS (rigid)	240	200	260	135
Acetal	205	256	220	218

should be less than 70 MPa (10 000 psi). Let us consider a moulding to be made from polycarbonate using a runner of length 100 mm and diameter 3 mm and with the melt at a temperature of 300°C. Let us also suppose that the volume flow rate (Q) of the plastic melt through the runner is $2.65\,cm^3/s$.

The first step is to calculate the shear rate (γ) using

$$\dot{\gamma} = \frac{4Q}{\pi r^3} = \frac{4 \times 2.65 \times 10^{-6}}{\pi \times (1.5 \times 10^{-3})^3}$$

which gives a shear rate of $1000\,s^{-1}$. (Note, all units must be in standard SI, hence $2.6\,cm^3/s = 2.6 \times 10^{-6}\,m^3/s$ and lengths must be converted to metres.)

At 300°C, a shear rate of $1000\,s^{-1}$ corresponds to a viscosity of 420 Pa s. (This information is obtained from the manufacturer's data for the particular grade of polycarbonate.) The shear stress can now be calculated using

$$\tau = \eta \dot{\gamma} = 420 \times 1000 = 0.42\,Pa$$

The pressure drop across the runner can be calculated using

$$P = 2\tau L/r = 2 \times 0.42 \times 100/1.5 = 56\,MPa$$

Let us now suppose a volume flow rate of $12\,cm^3/s$ is required. Using the same equations and procedure, it will be seen that

$$
\begin{aligned}
\text{shear rate} &= 4530\,s^{-1} \\
\text{corresponding viscosity} &= 200\,Pa\,s \\
\text{shear stress} &= 0.9\,MPa \\
\text{pressure drop} &= 120\,MPa
\end{aligned}
$$

This pressure drop is too high. It can be reduced by increasing the size of the runner. What is required is the diameter of the runner that will give a pressure drop of less than 70 MPa under the given injection conditions.

Increasing the runner diameter to 4 mm and repeating the above calculation procedure, the results are:

$$
\begin{aligned}
\text{shear rate} &= 1910\,\text{s}^{-1}\\
\text{corresponding viscosity} &= 325\,\text{Pa s}\\
\text{shear stress} &= 0.62\,\text{MPa}\\
\text{pressure drop} &= 62\,\text{MPa}
\end{aligned}
$$

This pressure drop is acceptable.

3.8   Channel cross-section

Cross-sections other than circular are used, the most common being trapezoidal and half-round, because of the ease with which they can be machined into one half of a mould. Although flow through such channels has been analysed, in practice it is found that if such a channel is used, there are dead areas of flow after a relatively short time. This is because the polymer flow in the corners of the section has stagnated and, ultimately, the plastic has degraded in many cases, leaving the central part of the channel as the flow channel (Figure 3.6). The flow path therefore becomes approximately circular in cross-section. It is therefore better to calculate flow for dimensioning on the basis of a circular channel which just fits inside the trapezoid, i.e. of radius equal to half the height of the trapezoid. This treatment is an approximation but is believed to give better and certainly acceptable results for dimensioning of runners in the long run. It is only valid, however, where the trapezoid can be regarded as being almost square otherwise one is dealing with an elliptical flow channel.

Finally, the shear rate calculated using the equation given here is the *apparent* shear rate since the formula is derived assuming Newtonian behaviour. The true shear rate can be obtained by applying the Rabinovitch correction or by using the method of representative flow to account for

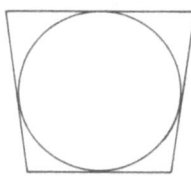

Figure 3.6 Showing circular flow path of polymer melt in a trapezoidal runner.

the discrepancy. In most cases, the use of apparent shear rate gives sufficient accuracy for runner dimensioning.

3.9   Filling the mould impression – other factors

Ultimately, the object of injection moulding is to fill the mould impression and then cool the moulding. These two requirements are not compatible in the commercial world. Ideally, the mould should remain hot (at the melt temperature) in order that the melt fluidity should be maintained at a constant level to facilitate mould filling. Commercial requirements are that the moulding should be cooled in as short a time as is possible while producing acceptable mouldings. When a hot melt is injected into a cold mould, cooling begins immediately and unevenly. Furthermore, the melt viscosity increases impeding flow and filling of the impression. Inevitably a compromise has to be reached in the interests of both satisfactory moulding and economic production. This inevitably leads to a number of problems as a result of less than ideal flow in the mould.

As the mould impression is filled, the material flows into the impression as depicted in Figure 3.7. The melt flows forward down the impression and sideways to the walls. The melt cools rapidly as it hits the walls and becomes immobile, forming an insulating layer for the central flow path. However, flow down the centre becomes increasingly difficult because the

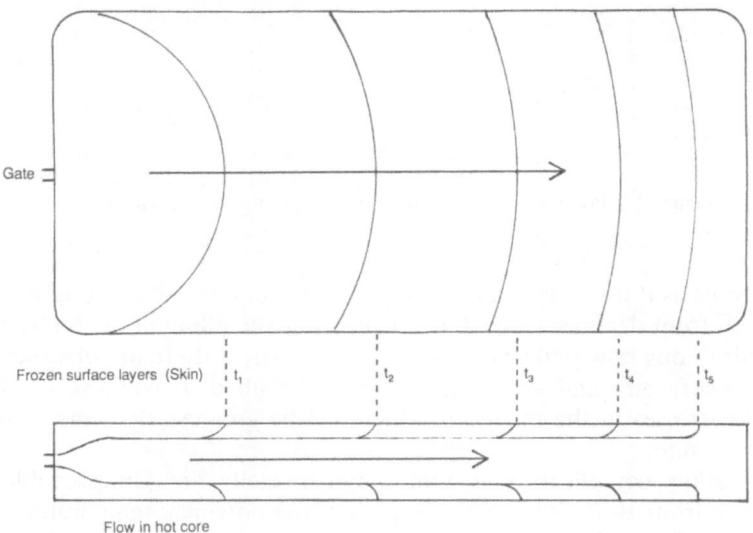

Figure 3.7 Showing schematically how the plastic melt flows into the impression.

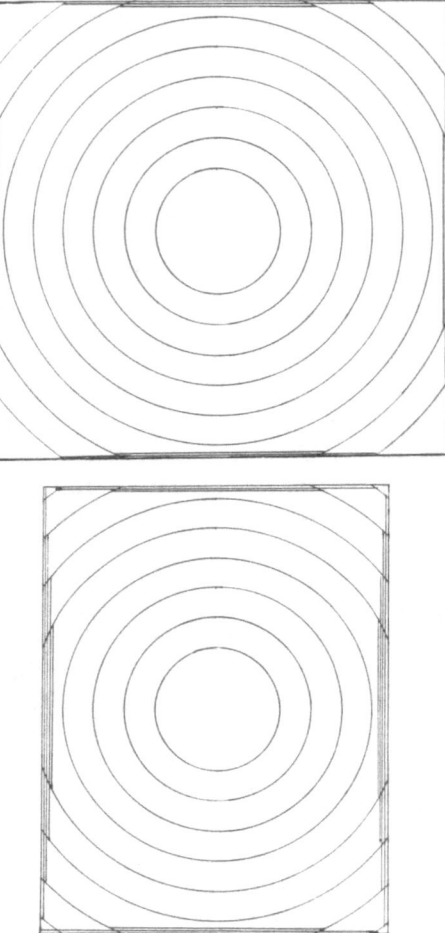

Figure 3.8 Problems of melt compressibility in injection moulding.

melt cools as it flows and in a given period of time the distance moved by the melt front decreases with time. This can make filling of the impression difficult if long flow paths are involved, especially if there are obstructions such as core pins and awkward corners to be filled. It will also result in built-in stresses in the moulding which will be greatest at points furthest from the gate.

A further complication is that polymer melts are compressible by anything from 10 to 30% depending upon the polymer, temperature and pressure. The implications of this may be seen by considering how the melt fills simple shape mould impressions. Suppose a circular dish mould is filled from a central gate. The flow front spreads out from the gate radially and the perimeter is reached by the flow front at the same time so that melt compression is uniform. If a thin square mould is fed from a central gate, then the melt will reach some parts of the perimeter before others (Figure 3.8). The melt will continue to fill the unfilled parts of the impression but the melt in contact with the perimeter cannot flow and will simply be compressed resulting in an uneven stress distribution across the moulding. The situation is similar with a rectangular mould but more complex.

Although moulding conditions can ameliorate these flow problems to some extent, the real answer is the use of correct gating and component design.

4 The injection moulding machine and its influence on mould design

4.1 Introduction

The process of injection moulding has today evolved to a high technical level. Major technical advances are not to be expected. Areas of interest to moulding machine designers and engineers include process control and quality assurance where inroads are still to be made as technology progresses.

The purpose of this chapter is to introduce the moulding machine to the would-be mould designer in terms of design and use requirement.

4.2 Machine function

Injection moulding machines are manufactured in many designs and configurations, often suited specifically to their intended application. Many contributing factors have to be taken into account before a mould can be designed to function in a given design of machine. In order to select a machine mould design combination and to realise the full potential of the process, it is necessary to consider the following factors:

(a) design of the proposed injection moulded component;
(b) the properties of the raw moulding material (resin);
(c) processing requirements in terms of component quality and production volume;
(d) economic viability.

The injection moulding machine can be broken down into three basic units (Figure 4.1):

(i)   the machine base unit;
(ii) the injection unit;
(iii) the clamp unit.

4.2.1 *The machine base unit*

Often referred to as the machine 'bed', this provides a mounting frame for the clamp and injection units. The prime functions of any machine

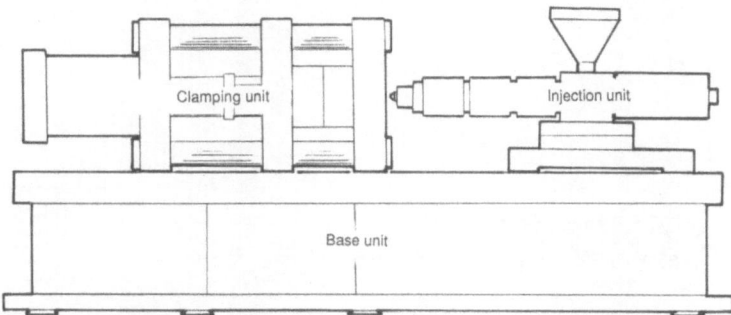

Figure 4.1 Injection moulding machine.

base unit must be dimensional stability, accuracy and strength. The process of injection moulding imparts high levels of stress to the unit during use which have to be absorbed without undue deformation occurring. The presence of high cyclic stress loadings during service dictates a rigid heavyweight construction of the unit. Modern machine base units are generally fabricated from heavyweight steel beams welded together by means of thick steel connecting side plates. Due to the relative ease of manufacturing method employed, the machine design can be readily modified during construction to fulfil specific requirements as necessary.

The base unit also acts as a housing for the hydraulic oil storage tanks. The pump and electric drive motor are often mounted on the back of the base frame or directly to the bottom of the structure.

4.2.2 *The injection unit*

The basic function of all machine injection units is that of:

- melting and preparation of the polymeric resin;
- pressurising and feeding the molten resin to the mould under controlled conditions.

Injection units are usually based upon two distinct designs of system, the plunger and the reciprocating screw designs. The latter is the most commonly employed on modern moulding machines. Figure 4.2 shows a typical reciprocating screw injection unit.

4.2.2.1 *Injection unit – basic operation.* Solid granules of resin are drawn into the screw flights as the screw is rotated by means of the screw drive motor. Heat is supplied to the resin through the barrel wall and additionally by friction generated as a result of the shearing action of the

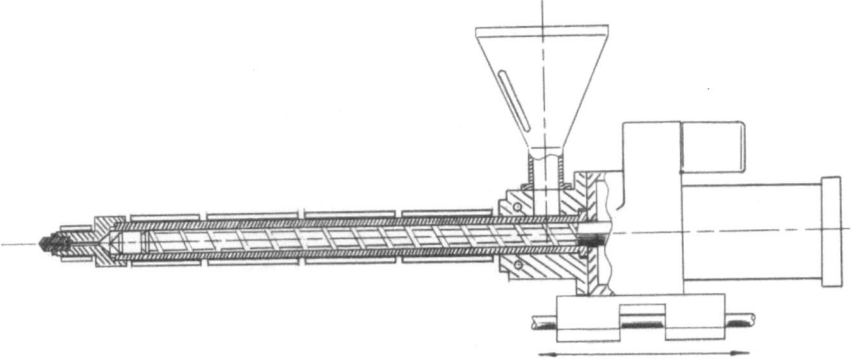

Figure 4.2 Typical reciprocating screw injection unit.

screw. The screw geometry creates a reduction in flight volume towards the front of the barrel thus compressing the softening resin into a molten phase. This process is often referred to as plasticising. The pressurised molten resin is then metered through a ring valve or 'collar' into the cavity between the screw tip and the nozzle. The build-up of resin in front of the screw tip is often referred to as the *shot* or *screw cushion*.

As the screw rotates, the screw cushion grows in size, forcing the screw to retreat back up the barrel until the required cushion size is achieved. Once this is obtained, screw rotation is halted. At this stage the injection unit is said to be *primed* and ready for the next operation, known as the injection phase.

To inject the molten resin through the nozzle into the mould, the screw is forced forward closing the ring valve in doing so. With the ring valve firmly seated shut against the screw shoulder, the screw cushion is thus pressurised. The cushion pressure builds until sufficient to force molten resin through the nozzle into the mould feed system. The applied pressure continues until the mould impression is completely full of molten resin. This sequence of events is referred to as the injection 'first stage' or high pressure phase. If the injection first stage is allowed to continue beyond the mould fill point, overpacking of the moulding will result. Overpacked mouldings suffer from high moulded-in stress levels resulting in poor component dimensional stability and inconsistent mechanical properties.

On completion of the first stage of injection, the applied pressure is reduced to a level sufficient to avoid component overpacking. Conversely, over reduction of pressure could result in component underpacking which leads to high component after-shrinkage and distortion. Known as holding, dwell, packing or second stage pressure, it is maintained until sufficient solidification of the feed system has occurred to allow removal of the applied pressure altogether. When the second stage has been completed, the injection unit can be reprimed in preparation for the next shot.

Metering control of the molten resin during the whole injection cycle is achieved by controlling the screw movement by adjusting the following variables in combinations appropriate to the machine:

- pressure applied to the screw;
- time allowed for movement;
- speed of movement;
- screw positional control.

The combined use of all or some of the above variables during the injection/holding sequence is often referred to as injection profiling.

4.2.3 *The clamp unit*

In order to inject the molten resin into the mould under considerable pressure, the mould must be sufficiently clamped together to resist the applied injection force. Under-clamping would result in the mould being forced apart creating flash about the split line. Apart from providing sufficient clamping force for the mould, the clamping unit is also utilised for mould opening, closing and sometimes component ejection when required.

Most modern injection moulding machines are hydraulically powered, the exceptions being the all-mechanical machines and more recently the all-electric machines at the smaller end of the market. The clamping force, although generated hydraulically, is transmitted mechanically by various means to the mould via the moving platen. Many designs of clamping system are used commercially; Figures 4.3 and 4.4 show diagrammatically two commonly encountered designs.

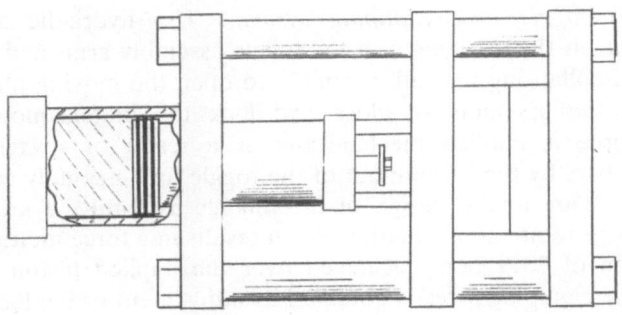

Figure 4.3 The direct hydraulic clamping system.

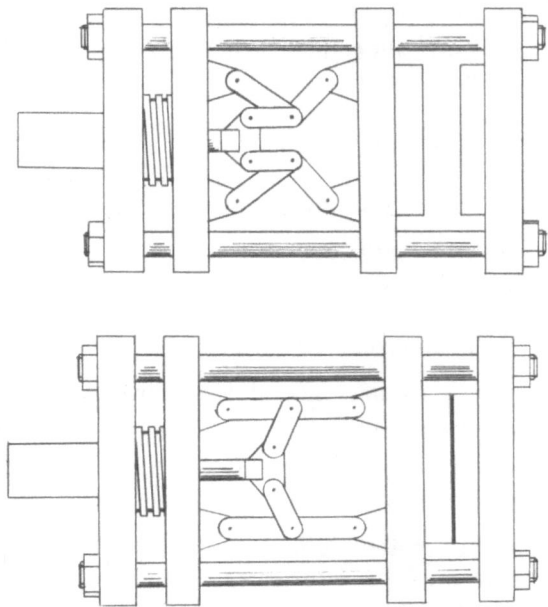

Figure 4.4 The toggle lock clamping system.

4.2.3.1 *The direct hydraulic clamping system.* Clamping force is generated by the application of hydraulic force on the ram piston which is directly attached to the machine moving platen. Platen forward movement and applied force is controlled by altering the fluid load pressure within the clamp cylinder. Applied pressure under the piston crown enables the platen to move in a backward motion to facilitate mould opening for ejection purposes. Platen movement speed control is often obtained by either throttling fluid input or output rates to the cylinder.

4.2.3.2 *The toggle lock clamping system.* The hydraulic actuation cylinder controls the movement of the toggle assembly arms and linkages by means of collapsing the link assembly to open the moving platen and erection of the assembly to close and lock the platens/mould. The clamping force is applied mechanically as a result of stretching the machine tie bars by the locking-out of the toggle link assembly contained within them. Due to the design of the linkage assembly, a mechanical advantage is generated on actuation which results in a force multiplication in the region of 20:1 being achieved over the applied piston loading. Setting of the clamping force is obtained by adjustment of the lock height which can be achieved by jacking the assembly forward or back on the lock jacking screw. Clamp tonnages can be calculated by measuring the resultant tie bar stretch generated by the subsequent locking down of the press after adjustment.

4.3 Machine types and configurations

The manner in which the injection moulding machine units are laid out and assembled constitutes the machine configuration. Factors which influence machine configuration choice include the type of material to be processed or the process to be undertaken. The two most commonly encountered machine configurations are the horizontal in-line and the vertical in-line designs. In each case the clamping unit directly faces the injection unit. Table 4.1 shows a number of machine design layouts often encountered.

Table 4.1 Machine design layouts

Machine type	Design configuration
a) Horizontal in-line	
Most commonly encountered design, allows gravitational free fall of ejected components away from mould halves on opening.	
Uses: general purpose moulding.	
b) Vertical in-line	
Requires less working floor space than horizontal in-line machine.	
Uses: insert loaded and over-moulded components, frequently used in conjunction with robots, etc.	
c) Horizontal lock – vertical inject	
Enables injection directly into the split line of the mould and free fall ejection to occur.	
Uses: enables direct feeding of components in smaller more economic bolsters.	
d) Vertical lock – horizontal inject	
Enables injection directly into the split line of the mould.	
Uses: frequently used for small insert loaded or multi-coloured overmoulded components.	
e) Multi-unit configurations	
Often manufactured directly to customer requirement (e.g. twin lock/inject configuration for the production of case halves).	

Company name   Customer name and address

Pt A General data

Work order no.:   Date:
Mould no.:   No. of impressions:
Material to be processed:
Component description:

Pt B Machine data

01 — Machine configuration   :
02 — Machine manufacturer   :
03 — Machine manufacturer's code   :
04 — Max. shot weight (grammes-PS)   :
05 — Max. clamping force (tonnes)   :
06 — Max. opening stroke (mm)   :
07 — Min. mould height (mm)   :

08 — Ejection   :
     Max. ejection stroke (mm)   :
     Type of ejection system   :
     Max. ejection force (kN)   :
     Ejector coupling thread size   :

09 — Platens   :
     Max. working tie bar clearance (mm) :
     Platen height (mm)   :
     Platen width (mm)   :
     Register ring diam. (mm)   :
     Please supply platen layout DRG   :   Yes/No (delete as appropriate)

10 — Injection unit   :
     Nozzle type   :
     Nozzle ball shut-off diam. (mm)   :
     No. of additional heater controllers :

Pt C Additional information

Figure 4.5 Machine specification data sheet (sample).

4.4 Machine specification

In addition to choosing the intended machine configuration, attention must be paid to the machine specification in order to enable the intended mould tool design to function efficiently. Failure to use an adequate machine specification during the mould design stage could lead to expensive tool modifications being necessary at a later stage. One of the most common mistakes encountered on a new mould tool is that the bolster is simply too large to fit the intended machine resulting in either tool modification or the tool working in a larger, more costly to run moulding machine.

Moulding machine manufacturers frequently specify their machine size ratings either by the lock tonnage available or the plasticising capacity of the injection unit (usually based upon polystyrene). For mould design use, the relevant machine data must be obtained and stored for reference purposes. The construction of a standard machine specification sheet (see Figure 4.5) can be of great help to the mould designer. Blank specification sheets can be sent to the proposed customer to fill in their machine details and returned to the tool designer on completion, thus reducing the likelihood of initial design error occurring. Such specification data sheets could also be used to constitute part of the tool order contract prior to commencement of work.

5   Understanding moulds

5.1   Introduction

When dealing with injection mould tools, toolmakers and engineers refer to the various mould component parts with a universally adopted terminology. Basic tool types and components can vary enormously both in shape and size, some having similar names, which can easily lead to confusion when encountered by people new to the trade. This chapter lists and explains the functions of the various individual mould parts which together make up the tool construction.

5.2   Tooling terminology

5.2.1   *The mould impression*

Often incorrectly referred to by some engineers as the cavity, the mould impression is a hollow area contained inside the mould into which molten polymer is injected and allowed to cool. When a mould tool contains more than a single impression, it is referred to as a multi-impression tool. Mould tool size is often characterised by the number of impressions contained within it, e.g. a 4-impression mould, 64-impression mould, etc. The impression is created as a result of bringing together two distinct components during closure of the mould, known as the core and cavity (Figure 5.1).

5.2.2   *The mould core*

The mould core forms the interior of the moulding and is recognisable as the male component which constitutes the impression. The core is retained or situated in a plate known as the core plate and the mould half which contains the core assembly is referred to as the core half of the mould tool. Due to the shrinkage characteristics encountered when polymer melts cool, the cooling moulding tends to shrink onto the mould core and away from the cavity wall once solidified. This characteristic shrinkage behaviour enables the moulding to be retained on the mould core during the mould opening sequence for ejection purposes. Since the machine ejection system is located behind the moving platen of the moulding

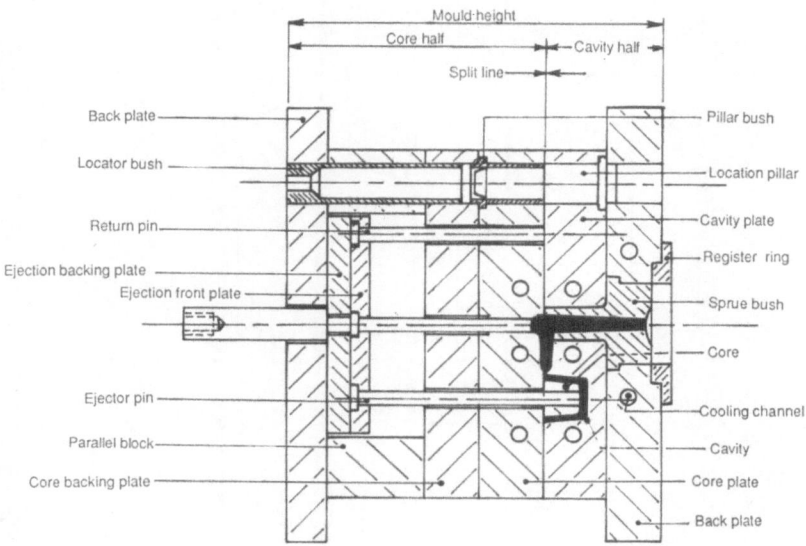

Figure 5.1 Basic mould tool.

machine (see chapter 4), the core half of the mould tool is usually clamped to the moving platen of the machine in use.

5.2.3   *The mould cavity*

The mould cavity forms the exterior or presentation face of the moulding and is recognisable as the female component which constitutes the impression. The cavity, like the core, is retained or situated in its own mould plate which bears its name, the cavity plate. The remaining half of the mould tool which contains the cavity is referred to as the cavity half and is usually clamped to the fixed platen of the moulding machine (see chapter 4) in front of the injection unit. The mould feed system is usually located in the cavity half of the mould due to its proximity to the machine injection unit.

With the tendency for tool designers to locate the ejection system in the core half and the feed system in the cavity half, it is easy to understand the influence tool design has over moulding machine choice of configuration. Hence the most commonly encountered machine design being the 'in-line' configuration (see chapter 4).

5.2.4   *The split line*

The point at which the core and cavity components interface on mould closure is referred to as the split or part line (Figure 5.1). Split line

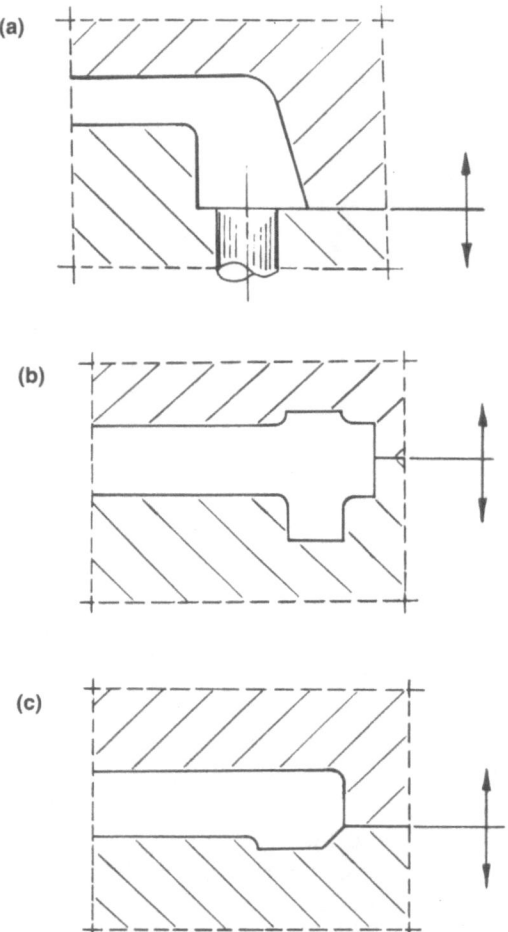

Figure 5.2 Examples of positioning of split line. (a) Split line deliberately off-set to simplify component ejection. (b) Split line moved up component side to simplify mould design and reduce toolmaking costs by removal of side core components from the design. (c) Positioning of the split line off the component frontal presentation face to disguise the presence of moulded flash.

position on the moulded component is determined by the complexity of the core and cavity interface. Mould designers often position the split line to simplify the mould design in terms of reducing tool sophistication and easing moulded component ejection. A few examples are shown in Figure 5.2.

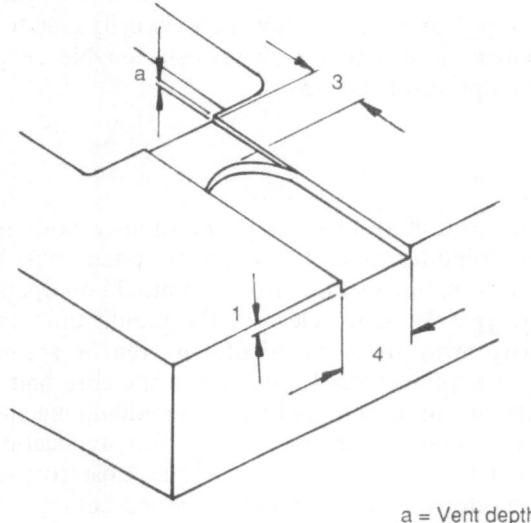

a = Vent depth

Figure 5.3 Venting.

5.2.5 *Venting*

During the cavity filling sequence of the moulding operation the incoming polymer melt forces trapped air into the mould. Due to the high tooling accuracy requirement of injection moulds, the trapped air is often unable to escape from the split line. For this reason, shallow slots (Figure 5.3) are ground into the split line mating faces to allow the trapped air to escape during mould filling. The slot depths are sufficient to allow air to escape but not deep enough to allow molten polymer to flow down them and create flash. The individual slots are referred to as 'vents' and collectively as 'venting'.

Table 5.1 Vent depth requirements of some commonly encountered polymers

Material	Maximum recommended vent depth
Nylon 6,6	0.0004/0.0006 in (0.01/0.015 mm)
Acetal	0.0006/0.0008 in (0.015/0.02 mm)
HDPE	0.0006/0.0008 in (0.015/0.02 mm)
PP	0.0007/0.0010 in (0.018/0.025 mm)
PS	0.0012/0.0015 in (0.042/0.06 mm)
ABS	0.0012/0.0015 in (0.042/0.06 mm)
Acrylic	0.0015/0.0020 in (0.06/0.078 mm)
PC	0.0015/0.0020 in (0.06/0.078 mm)

Note: All data refers to GP resin grades only.

Vent slot depth is determined by the viscosity of the polymer resin to be moulded; polymers of very low melt viscosity dictate shallow vent slots whereas polymers of higher melt viscosity enable deeper more efficient slots to be employed (Table 5.1)

5.2.6 *Ejection*

Mould tool ejection systems vary enormously both in complexity and design. The prime function of any ejector system must be that of clearing the moulded components away from the mould on opening, thus enabling the press to recycle. Once clear of the mould the components *free fall* under gravity away from the moulding area or are removed by other means, e.g. by robot. Usually situated in the core half of the mould, the ejection system can be actuated by the moulding machine or the opening action of the mould. Ejection design and application is discussed in greater detail in chapter 9. A few of the most commonly encountered ejection components are listed and described below.

(a) *Ejector bar*: The ejector bar mechanically interfaces the mould ejection system with that of the moulding machine. The bar can either be linked to the mould (e.g. by a screw thread) or used as a *knocker bar* without any mechanical linkage between mould or machine.

(b) *Ejector plates*: The ejector bar transmits the ejection force as a single point load to the centre of the ejector plates (Figure 5.1). The applied load is distributed to the various components attached or contained within the plate assembly. Ejector plates must be of rigid construction in order to withstand the relatively high cyclic loads imposed on them during service. If the plates flex or bow under load, tool wear may occur which could significantly shorten the mould's service life.

(c) *Support pillars and parallel blocks*: These components transmit the clamp force from the mould back plate through and about the ejector plate recess directly to the core plate. The support pillars additionally act as guides to locate the ejector plate assembly. The amount of ejection stroke available is determined primarily by the overall height of these two components.

(d) *Ejector pins and blades*: These headed components are trapped between the ejector plates. The ejector pins and blades are usually positioned within the core assembly, seated in reamed holes and ground flush to the face of the core with the ejector plates fully back (Figure 5.1). Components such as these are subject to considerable wear in use due to the continual rubbing action within the core. Most pins and blades fitted to modern mould tools are standard components, often mass produced and catalogue ordered for use. Ejector pins and blades are good examples of the application of standardisation in reducing tool maintainance and construction costs (chapter 13).

(e) *Stripper plate*: The use of a stripper plate enables the applied ejection force to be transmitted to the periphery of the moulded component resulting in a more supportive ejection method than either pins or blades. They are frequently employed on thin-walled mouldings to avoid damage. The stripper plate is positioned in front of the core plate over the extended core bodies which also serve to locate the plate, usually on a taper fit. The stripper plate is guided on the mould's main location pillars. The ejection force can be applied to the plate in a number of ways, e.g. (i) pulled by the opening action of the press; (ii) pushed directly by the ejector bar; (iii) pushed indirectly by the ejector plates via push rods.

5.2.7 *Back plate*

The function of the mould back plates is often overlooked by mould designers who tend to focus on the core and cavity build-up. The split line constitutes the moulding datum whose accuracy is dependent upon the core and cavity mould half build-ups, which in turn rely on the back plates as working datums. Apart from their use as working datums, the back plates are frequently used for the following functions:

(a) as bolting points to hold the individual tool halves together;
(b) to provide securing points to either clamp or bolt the mould into the press;
(c) to provide rigidity and integrity to the construction of the mould;
(d) to standardise bolster dimensions from one mould to another (chapter 13).

5.2.8 *Sprue bush*

The sprue bush is a headed cylindrical component (Figure 5.1) about a tapered polished bore. The bush is usually positioned in the cavity half of the mould and provides an entry point for the machine injection unit to feed the mould. To minimise the chance of leakage between the bush and the machine nozzle, a spherical concave radius is often machined into the head of the component to provide a suitable contact seat. The feed passageway is polished and tapered to ensure the solidified sprue is pulled back with the core half of the tool on mould opening for ejection purposes. Sprue bushes are frequently damaged in service and should be treated as a standard mould part by mould designers.

5.2.9   *Register ring*

This is a disc-shaped component seated into a recess in the mould back plate on the cavity half of the tool. Its function is that of locating the mould on the fixed platen of the press. The register ring, being an accurate component, ensures the sprue bush aligns on the centre line of the injection unit thus reducing the likelihood of nozzle leakage occurring during use. The register ring can additionally be employed to retain the sprue bush (Figure 5.1) if desired. A badly aligned register ring could lead to nozzle leakage during use which could cause the sprue to stick in the sprue bush on mould opening or, worse still, damage the machine nozzle as well as the sprue bush. Register rings, like sprue bushes, ejector pins, etc. should also be treated as standard mould components by mould designers.

5.2.10   *Tool location*

Provision for adequate location of the mould tool must be made to ensure the integrity and accuracy of the moulded component. Mould location can be divided into two categories, primary location and secondary location. Primary location usually takes the form of pillars and bushes (Figure 5.1) positioned in the corners of the core and cavity plates respectively. Secondary location is additionally provided when the component accuracy requirement demands it, i.e. the moulding of thin-walled components. Taper location bushes or crossed location blocks are frequently utilised for this purpose and are usually positioned on the centre line of the mould halves for maximum effect. Mould tool location is subject to wear during tool use and should be considered as part of the standard component category by tool designers and toolmakers alike.

5.3   Mould types

Injection mould design types and configurations vary enormously due to the many technical requirements of the moulded components to be produced. For this reason the categorisation of mould tool type can only be generally applied. Of all the mould configurations employed, the two-plate and three-plate mould designs are the most easily recognisable in appearance. The two- and three-plate designs owe their names to the number of split or parting lines (Figure 5.4) employed within the tool design. Figure 5.4 illustrates the significance of split line position with respect to the individual mould designs.

The two- and three-plate designs form the basis for many variations on their basic design format. The following are examples to illustrate how they may be employed:

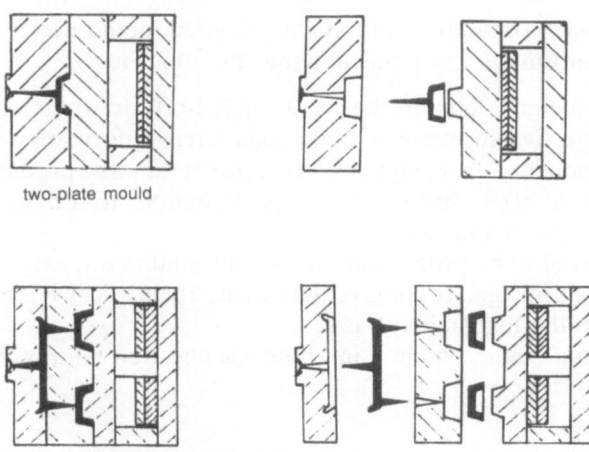

two-plate mould

three-plate mould

Figure 5.4 Significance of split line position with respect to individual mould designs.

(a) Runnerless moulds, more frequently referred to as hot runner moulds, in which the material feed system remains molten throughout the entire moulding cycle, only freezing off at the gate of the moulding prior to ejection from the mould (see chapter 11).

(b) Undercut moulds, which produce components whose features undercut the mould cavity or core components in such a manner as to require core pulling or feature removal prior to mould opening or ejection (see chapter 12).

(c) Family moulds, in which many different components are manufactured within the same tool, e.g. model kits, kitchen utensils, etc. Family moulds are often employed on short run, relatively low volume ranges of components to reduce tooling costs.

(d) Stack moulds in which the impressions are stacked either side of the mould centre plates, often used on vertical moulding machines for compression moulding purposes, e.g. the manufacture of insert loaded components, records, sinter moulded components made in PTFE, etc. (see chapter 10).

5.4 Choosing the correct mould

Choice of mould design and construction method is determined by the number of factors and requirements imposed on the tool designer. Mistakes can easily be made if all the relevant factors and requirements are not obtained and considered beforehand. A good mould designer will obtain as much information as possible about the following:

(a) Component: obtain the final signed-off drawing issue from the prospective customer. Obtain commercial information regarding the component, e.g. component cost, size of required production batches, estimate of production cycle time acceptable to the customer, mould life requirement, etc.
(b) Material to be processed: obtain information on material properties, e.g. shrinkage, cooling requirement, rheological features (chapters 1–3), thermal stability, etc.
(c) Machine data: obtain a machine specification (section 4.4).

                                                      Form no.
                        Mould specification form

Comp. description:
Mould number: Date:
Customer name and address:

Pt A Component information

Comp. drg no.   Issue
etc.

Pt B Material to be processed

Material description
Material manufacturer   Grade no.
etc.

Pt C Machine data

Data sheet supplied   YES/NO   (delete as appropriate)
Machine data sheet no.   Sheet attached   YES/NO
etc.

Pt D Customer preferred data

(e.g. type of hot runner system, cooling fixtures and fittings, eyebolt sizes, standard clamping heights, etc.)

Pt E Additional comments
(e.g. problems with similar components, etc.)

Figure 5.5

Once obtained, the relevant information should be used to construct the mould specification for future use and reference.

5.5 Mould specification

The use of a mould specification has a number of advantages to the mould designer in terms of reducing the likelihood of unnecessary mistakes occurring and building up the designer's background knowledge of related subjects. A good mould specification also provides the mould designer with the basis for a quality control system for customer audit or approval requirement thus increasing customer confidence, a factor of importance with respect to further order placements.

The easiest way of compiling a mould specification is by the use of a standard formatted form with all the required information laid out in a logical sequence. Figure 5.5 suggests a possible layout for such a form.

6 The two-plate mould

6.1 Introduction

Two-plate moulds are the most common design of injection mould tool used in the moulding industry. Mould designers choose the two-plate format because it offers many advantages in terms of simplicity of design, user friendliness, utilisation of standard mould parts and, above all, it often represents the cheapest design option available. The main disadvantages of the two-plate mould design are limitations in component gate positioning when conventionally feeding, lack of available space for balanced feeding of multiple cavities and high material waste levels (sprues and runners).

6.2 Tool construction

The two-plate mould is the simplest of all the mould design configurations, being constructed from two distinct half units, the core half and the cavity half. The point at which the two halves interface is known as the split or part line which divides on opening of the mould for component ejection purposes. The core half of the mould is usually attached to the moving platen of the moulding machine since the mould ejection actuation system (usually in the form of a hydraulic cylinder) is commonly positioned behind the moving platen on the moulding machine. The mould ejection system is correspondingly built into the core half of the mould for actuation purposes. The cavity half of the mould is therefore attached to the fixed platen of the moulding machine directly in front of the machine injection unit for material feeding of the mould (chapter 4). Cooling channels are positioned in both core and cavity components to control the mould temperature during use (chapter 8).

There are various methods by which the core and cavity may be incorporated into their respective halves of the mould. The two most commonly encountered are:

(a) The integral or integer method which involves machining the core or cavity form directly into the core or cavity plates respectively (Figure 6.2).

(b) The inserted bolster method in which the core and cavity is built up

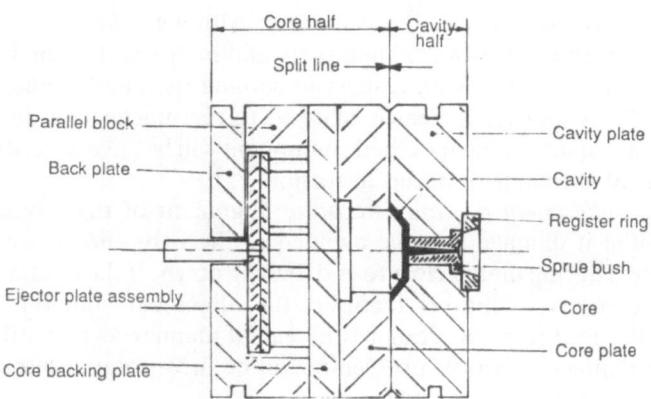

Figure 6.1 The two-plate mould.

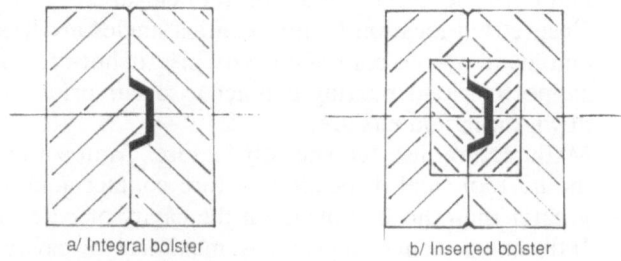

a/ Integral bolster   b/ Inserted bolster

Figure 6.2 Core and cavity incorporation techniques.

from individually machined components and secured into machined pockets situated in the core and cavity plates (Figure 6.2).

The choice of mould construction method must be carefully considered by the mould designer who must bear in mind the following points:

(a) *Complexity of the core/cavity build-up.* Detailed and complex core or cavity build-ups cannot realistically be accommodated within the integer method of mould construction. Pocketed location of core or cavity inserts offers advantages in terms of mould serviceability and achieving accuracy requirements.

(b) *Accuracy requirement of the final end product.* In order to achieve a good standard of component accuracy, the tool design should allow for the individual machining of core and cavity component detail. The use of core or cavity inserts enables the tool maker to realistically achieve tooling accuracy and detail requirements in terms of positional accuracy and form.

(c) *Cooling requirement of the moulding.* Although the use of inserted detail often reduces machining costs and increases tooling accuracy, problems may arise with respect to cooling the mould generally and the inserts locally. If heat is allowed to rise unchecked during tool use, the quality of the moulded component will be adversely affected in terms of shrinkage-induced distortion.

(d) *Mould alignment method.* Accurate alignment of the mould tool is essential if damage is to be avoided to the core and cavity shut-off features during mould closure and locking down. If the mould contains fragile inserted shut-off features, the alignment components built into the tool must be designed in such a manner as to protect them from damage. Shut-off protection can be incorporated into a mould design in the following ways:

    (i)   Ensure that the location pillars and bushes or shut-offs (heal blocks) engage the opposite side of the mould before the core and cavity features contact on tool closure.

    (ii)   Ensure that location feature contact angles are less than those employed on the core and cavity insert shut-offs. This enables supportive load bearing contact to occur prior to insert lock down on mould closure.

    (iii)   Manufacture inserted shut-off features from softer, less brittle mould tool steel if location feature contact angles have to be greater than those allowed on the cavity or core inserts.

    (iv)   If the component design allows, maximise the cavity draft angle, especially when hard or rigid polymers (e.g. polycarbonate) are to be moulded. Increasing the cavity draft angle reduces the likelihood of mouldings sticking in the cavity on opening of the mould. Stuck mouldings will damage core/cavity shut-offs during closure of the mould.

(e) *Size of mould.* Large mould core and cavity plates are often too big to fit on conventional machine tools (e.g. milling machines, surface grinders, etc.), especially if detailed machining work has to be undertaken. The use of an inserted bolster tool design is advisable if expensive machining costs are to be avoided or if there is a likelihood of scrap occurring during manufacture.

(f) *Heat treatment requirement.* Mould core and cavity components often require heat treatment to render them hard enough to avoid wear and damage while in service. As a rule, whole plates are not usually hardened. Therefore if core and cavity hardness is a service requirement of the mould design the inserted bolster tool construction method is employed.

(g) *Mould tool serviceability requirement.* The servicing of production quality moulds is often a costly exercise, both in terms of mould down time (lost production) and bench costs (fitting labour and part costs incurred). Moulds can be designed to aid servicing and reduce downtime by inclusion of the following design features:

(i)   Utilisation of standard mould parts (see chapter 13) in high wear areas, e.g. ejector pins and bushes, locator pillars and bushes, sprue bushes, standard heater sizes (see chapter 11), etc.

(ii)   Incorporation of high wear core and cavity features as inserts, e.g. shut-off faces. Gates can be carried within individual inserts.

(iii) Incorporation of frontal access to wear features, e.g. bolt or secure core and cavity features from the front of the plate faces. This enables the removal of these features from the mould without taking the mould out of the moulding machine.

(iv) Simplification of the design of wear features in terms of component complexity and accuracy.

(h) *Design flexibility.* The manufacture of certain products, e.g. bottle caps and closures, can demand a high degree of design flexibility from the mould tool. The changing of core and cavity components during production, often in the moulding machine, is a common requirement occurrence in the packaging industry or when prototyping new components (see chapter 14). Mould designs should therefore be easily accessible and relatively simple in design construction to minimise labour requirement and down time; such mould designs will be covered in chapter 14 in greater detail.

(i) *Financial constraints imposed upon project.* The financial constraints, in terms of project funding, must be fully understood by the mould design engineer prior to starting the project. The mould design must take into account the projected mould life, usually with respect to the component to be produced. For example, moulds of long life requirement will require hardened core and cavity components; this will incur greater expense in terms of machining and component finishing. A mould of short life requirement may not realistically require hardened core and cavity components, thus incurring lesser production expenditure. Money is often wasted when moulds are designed and manufactured in excess of use requirements, conversely under engineering of a mould tool will prove costly if mould parts wear out and have to be replaced before time.

7 Runner and gate design

7.1 Introduction

The feeding of the mould impression is of great importance to the mould design engineer. The geometry, length, size and volume of the feed system has a direct influence on the quality and integrity of the moulded component. The path by which the molten polymer flows through the feed system should always be the smoothest and least tortuous possible, over the shortest possible flow distance.

A basic mould feed system comprises three distinct components: the sprue, the runner and the gate (Figure 7.1).

7.2 Freeze flow characteristics

To design a mould feed system effectively the design engineer must understand the flow nature of polymer melts and the influences that flow channel geometry impose upon pressure transmission and flow efficiency. When a hot molten polymer comes into contact with a cooler metal mould, such as a runner, the surface of the polymer freezes on contact to form a skin. The resultant frozen skin acts to thermally insulate the molten core of the remaining melt flow (chapter 2). The geometry oof the flow channel tends to dictate the shape of the frozen skin about the melt flow, resulting in a pressure drop occurring proportional to the drop in effective flow volume. Known as freeze flow patterns, some commonly encountered flow channel sections and their resultant freeze flow patterns are illustrated in Figure 7.2.

The freeze flow characteristics created by the runner geometry have a marked effect on channel flow rate and pressure transmission. Section profiles which exhibit flat surfaces or tend to be angular in form, e.g. square, rectangular and semi-circular, suffer from poor freeze flow characteristics in use. The resultant reduction in effective flow volume within such channel sections increases the pressure drop within the channel, thus reducing cavity pressure control within the moulding.

The section profile of the runner also affects the ejection force requirement of the feed system. Runners of curved or angular form, e.g. half/full round or trapezoidal sections, require significantly less ejection force to eject them, compared with square or rectangular runners. Runner sections

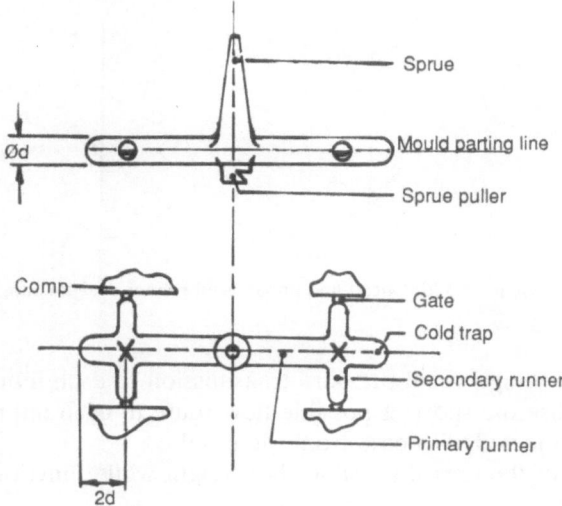

Figure 7.1 Basic mould feed system.

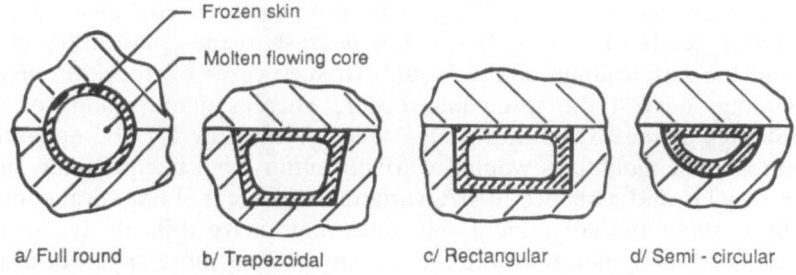

Figure 7.2 Commonly encountered flow sections and their resultant freeze flow patterns.

of the full round form offer the best flow and ejection characteristics of all the runner designs, but being machined into both halves of the mould they are the most expensive to produce. To reduce the need for full round runners, the curved base trapezoidal runner form is employed offering a compromise between cheaper machining costs and acceptable flow and ejection characteristics. A special purpose milling cutter is required to machine the runner form into one half of the mould during manufacture.

7.3 Runner configurations

The runner design layout or configuration chosen by the mould designer should exhibit the following attributes:

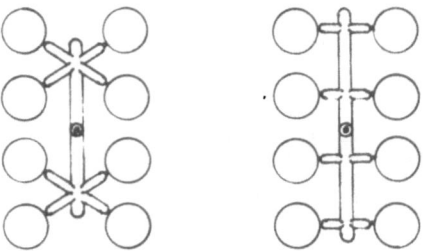

Figure 7.3 Balanced and unbalanced runner design configurations.

(a)   encourage equal pressure transmission to each mould impression;
(b)   utilise the shortest possible flow route to each impression;
(c)   be ejected with ease from the mould;
(d)   be of the lowest possible shot weight while functioning adequately.

Two basic design groups of runner configuration exist: one of balanced pressure transfer design, the other of unbalanced design (Figure 7.3).

The choice of configuration to adopt is usually dependent on the required quality of the moulding to be produced. The unbalanced configuration tends to favour the cavities nearest to the sprue entry point during the initial filling of the mould. In such cases overpacking of the mouldings nearest the sprue may occur, whereas underpacking of the mouldings at the extremities of the feed system may also be apparent. The resultant mouldings would invariably suffer from unequal individual shot weights and a chance of size variation will occur. Long-term control of flash about the component split lines may prove difficult due to the unbalanced filling nature of the tool which becomes more apparent as the tool beds in with age. In the case of a mould which exhibits an unbalanced filling profile, the individual cavity pressure drop can be compensated for by altering the individual gate dimensions, or probe position in the case of runnerless moulds (chapter 11), until a filling balance is achieved.

With respect to all runner system designs, all runner branch junctions or intersections should be curved and blended with radii where possible in order to minimise the pressure drop induced by the flow of the molten polymer through the system. The addition of cold traps (Figure 7.3) at runner end junctions serve to reduce the clogging effects of chilled slugs of polymer formed at the front of the melt flow as it progresses through the system.

7.4   Gate positioning and design

The location of the gate is of importance with respect to the way in which the polymer flows into the mould impression. Other factors have to be taken into consideration when choosing a gate location, which include:

(a) aesthetic considerations of the moulding;
(b) degating requirements of the moulding;
(c) design complexity of the moulding;
(d) mould temperature requirement;
(e) nature of the polymer to be processed;
(f) volume of the polymer to be fed though the gate and the feed rate;
(g) significance and positions of weld lines produced;
(h) possible locations and effects of gas entrapments created as a result of the filling profile.

After consideration of all the previous points, it must be apparent that the positioning of the gate must be determined relatively early on and before the design build-up has started. If the component is to be conventionally fed, i.e. non-hot runner, the gate location options are further restricted due to the limitations of the two-plate mould design. Restrictions exist because of the necessity of having to gate on a single plane about the mould split line. When confronted with gating requirements outside the scope of the two-plate design, the mould designer has the choice of selecting either the three-plate (chapter 10) or the runnerless mould (chapter 11) design formats in order to achieve the desired gate position.

7.5 Gate types

Once the most probable gate location has been decided, a gate of the correct design and geometry has to be chosen. Gate designs differ according to their intended application. The factors which affect the choice of gate location also affect the choice of gate design to be adopted. For single impression moulds or individually large components, a directly fed gating system can be employed (Figure 7.4). In such cases an extended machine nozzle feeds directly onto the mould impression, thus removing the need for a runner system. To achieve an even fill of the moulding without weld lines, directly fed gates are usually situated at the geometric centre of the component, e.g. dustbin bodies, buckets, bowls, etc., when possible.

7.6 Edge gate

The edge gate (Figure 7.5) represents the simplest of gate designs, being the easiest to produce, often by one pass of a milling cutter. Although easily produced, the edge gate is limited in use, primarily due to the poor filling characteristics of the design. The gate geometry dictates a poor spread pattern upon the exiting melt flow. In the worst cases, when the

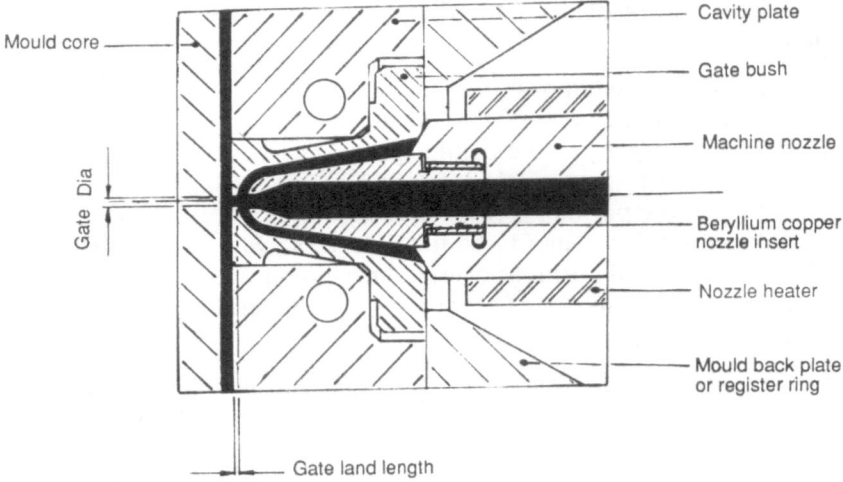

Figure 7.4 Direct component gating.

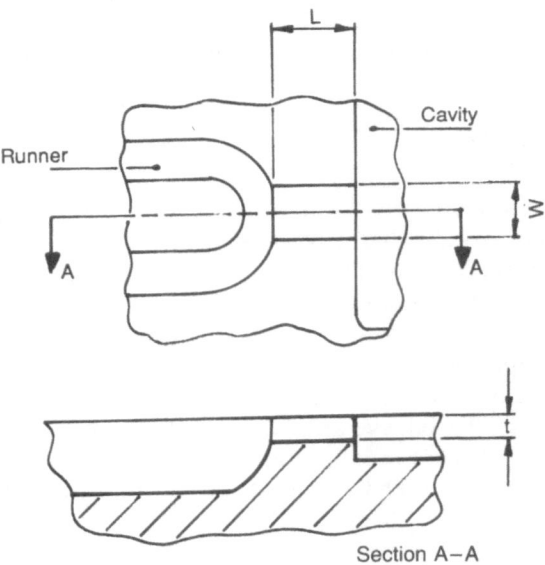

Figure 7.5 Edge gate geometry. W, width of gate; influences the melt flow pattern on exit from the gate. L, gate land length; influences the pressure drop created by the presence of the gate. t, depth of gate; influences the volumetric throughput of the gate.

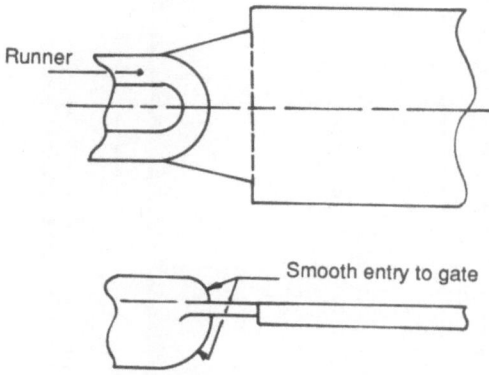

Figure 7.6 Typical fan gate.

exit pattern fails to spread at all, a twisting wormlike output may result known as worming or jetting. In such cases the resultant moulding suffers from poor integral strength and surface finish. Edge gates are commonly employed to fill mouldings of reduced quality requirement or aesthetic appearance. A degating operation is additionally required resulting in an unsightly gate scar.

7.7 Fan gate

Fan gates (Figure 7.6) are essentially an expanded design of edge gate, used to feed thin sections. The molten flow of material fanning out from the gate results in a uniform filling effect. Fan gating helps to reduce component warpage and improve component surface finish. For effective results, the contact area of the gate should never exceed the cross-sectional area of the feed runner. Due to the large gating area, unsightly degating marks can be a problem after runner removal.

7.8 Diaphragm gate

Diaphragm or disc gating (Figure 7.7) can be used for cylindrical or hollow section components when concentricity and weld strength are component function requirements. To balance mould filling, a minimum gate land length of 0.5 to 1.0 mm is generally recommended. The subsequent removal of the gate disc from the moulding tends to leave a sharp jagged edge about the rim of the component. Positioning of the gate inside the moulding serves to reduce post-mould finishing operations.

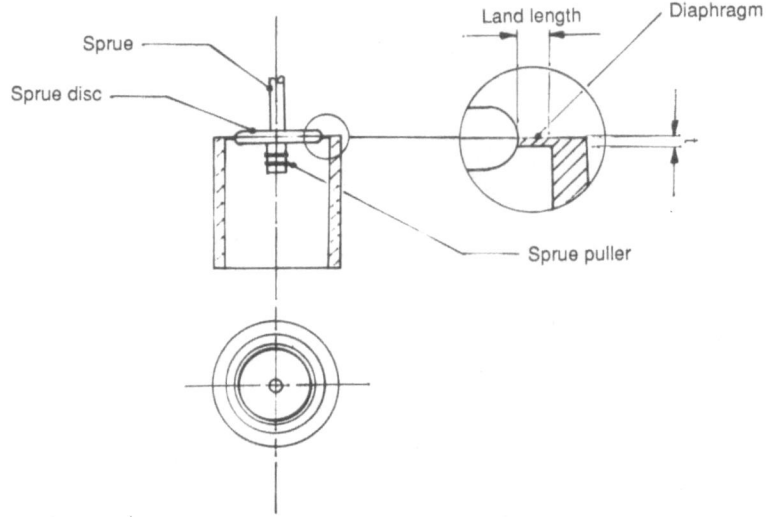

Figure 7.7 Diaphragm gate.

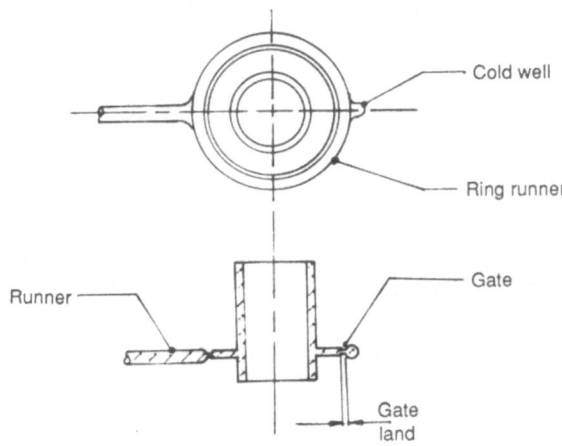

Figure 7.8 Ring gate.

7.9   Ring gate

Ring gates (Figure 7.8) are commonly employed on cylindrical mouldings when internal dimensions are more crucial than external. A runner 'gutter' is machined about the mould cavity and subsequently mirrored

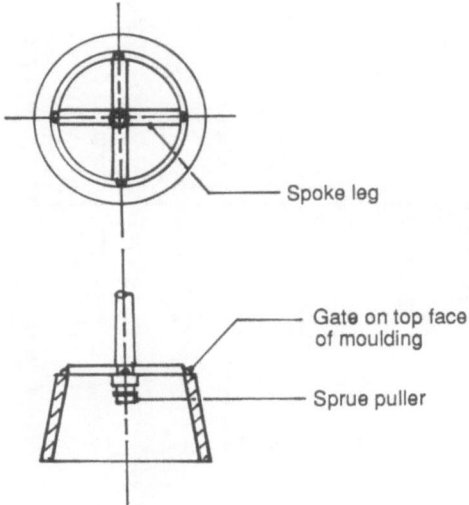

Spoke leg

Gate on top face
of moulding

Sprue puller

Figure 7.9 Spoke gate.

about the core, and a land is machined connecting the gutter to the cavity wall. Gate land depth is usually determined as a result of mould trials, shallow lands detract from the packing control of the moulding, whereas over-thick lands create degating problems and can lead to increased costs.

7.10 Spoke gate

Spoke gates (Figure 7.9) can be employed on mouldings which are often too large to diaphragm or ring gate. They allow a larger volume of polymer to flow through them than either of the latter two designs. In this case component accuracy and weld strength is reduced in favour of volumetric throughput and packing control. This design of gate is suitable for thicker section cylindrically shaped mouldings with high shot weights. The large gate legs usually require a machining operation to remove them after moulding.

7.11 Tunnel or submarine gate

Tunnel gating (Figure 7.10) permits the automatic degating of the moulding from the feed system. The gate is sheared off the component during the ejection cycle of the moulding process. Tunnel gate diameters

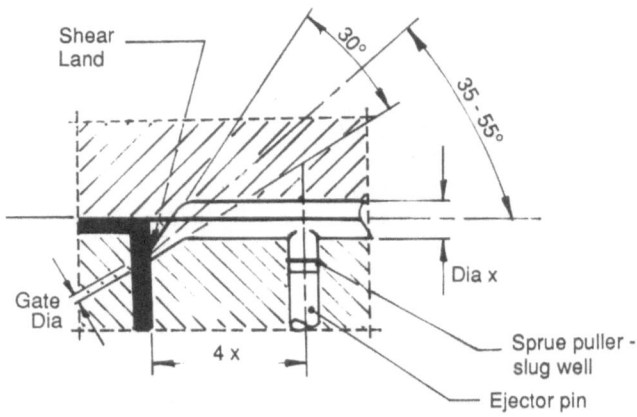

Figure 7.10 Tunnel gate.

vary from 0.5–0.8 mm for unreinforced plastics to larger diameters of 2 mm plus for reinforced materials. Due to the buried design of the gate, gas trapping and the resultant burning of the moulding surface can prove a problem in use. Bearing this in mind, adequate venting must be added to a mould which incorporates a tunnel gating design. A gate scar is left on the surface of the moulding, which subsequently increases in size as the gate shear land wears.

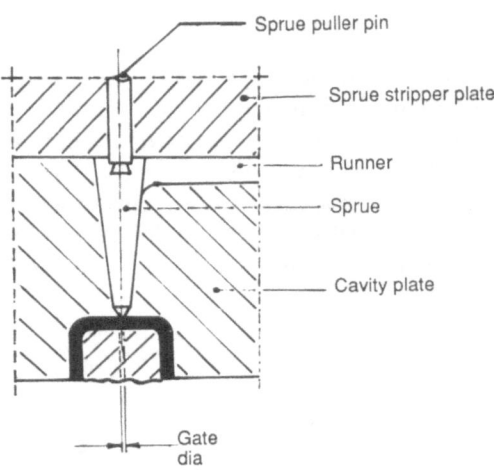

Figure 7.11 Pin point gate.

7.12 Pin point gate

Pin point gates for three-plate moulds vary in size from diameters of 0.8 mm–2 mm for unloaded materials to diameters of 2.5–3 mm for loaded grades. This design of gate also permits the automatic degating of the mouldings during use. The gate land is usually reduced in size to enable a 'clean' break to occur on separation from the moulded component. Gate breakage is achieved by means of pulling the runner from behind, usually by sucker or puller pins buried in the back of the runner (see Figure 7.11). As a result of reduced gate land thickness, pin point gates are easily damaged and tend to suffer from wear if loaded materials are processed regularly.

7.13 Tab gate

The side filling effect of the tab gate format (Figure 7.12) reduces the likelihood of jetting or worming occurring when the melt flow exits the gate. Tab gates are frequently employed to gate large decorative mouldings such as housings or instrument cases for aesthetic reasons. The even fill pattern produced about the tab feature serves to reduce the effects of stress-induced distortion or warpage once the moulding has solidified. Tab features are expensive to remove from mouldings and should therefore be positioned in locations where they can be left on the component or have a use function, e.g. as holding tabs for paint spraying or similar finishing purposes.

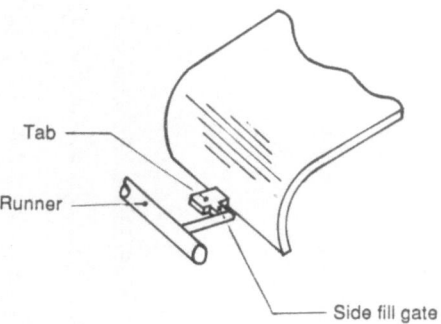

Figure 7.12 Tab gate.

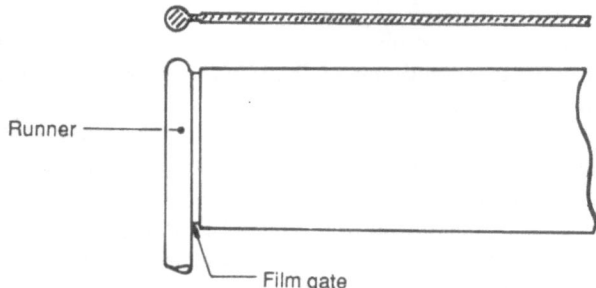

Figure 7.13 Flash or film gate.

7.14   Flash or film gate

Flash gates (Figure 7.13) are essentially an extension of the fan gate
principle. They can be used on mouldings of flat design or on large areas
where warpage must be minimised. Flash gates are expensive to deploy
due to their high removal and subsequent component finishing costs. The
end use requirement of the moulded component should therefore justify
the selection and use of such a gate design.

8 Mould cooling

8.1 Introduction

The importance of adequate mould cooling cannot be overstated. Failure to maintain a balanced mould temperature during use affects the dimensional stability and integrity of the moulded component. As mould temperature rises, cooling times are increased to compensate for the reduction in cooling efficiency. This produces a rise in production costs per unit produced.

8.2 Cooling requirement

The thermal properties of individual polymers differ markedly from each other (chapter 1), especially in terms of melt flow temperature and specific heat capacity. Bearing this in mind the design of the mould cooling system should reflect this in some way by compensating for the individual requirements of the subject polymer. For example, polypropylene has a significantly higher specific heat capacity than, say, polystyrene and therefore has to be treated accordingly (e.g. increase the coolant flow rate through the cooling system).

During the injection moulding cycle, almost 75% of the total cycle time (Figure 8.1) can be spent cooling the moulded component. This means that the press remains largely idle up to the point of opening and ejection.

To effectively cool the moulding, the mould cooling channels have to be accurately positioned. The cooling effect must be greatest where the moulding is the hottest and progressively less effective where the moulding is cooler. If the mould designer is able to achieve a balanced thermal state within the mould/moulding this will be reflected in improved component quality.

Figure 8.2 illustrates the thermal profile or thermal footprint of a typical moulded component of circular design, being centrally gated from behind. With respect to Figure 8.2, the hottest area can be found at the centre of the moulding within the immediate vicinity of the gate. The cooler areas are to be found on or near the edges of the component where the polymeric flow is reduced during cavity filling. To obtain a moulding with minimal warpage and stress, the component requires maximum heat removal from its centre and reduced heat removal from its

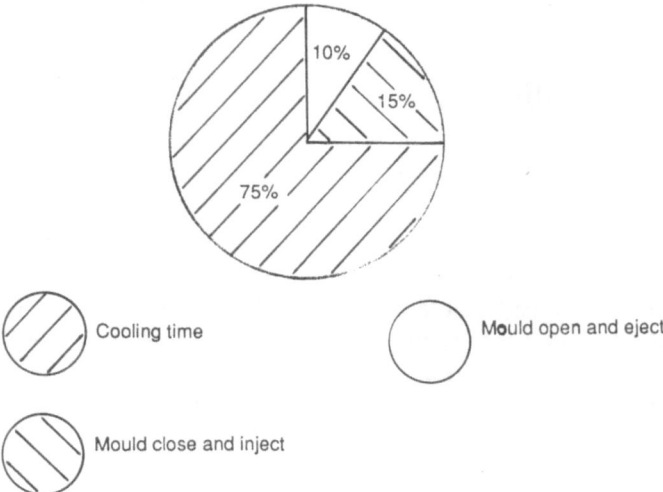

Figure 8.1 The moulding cycle.

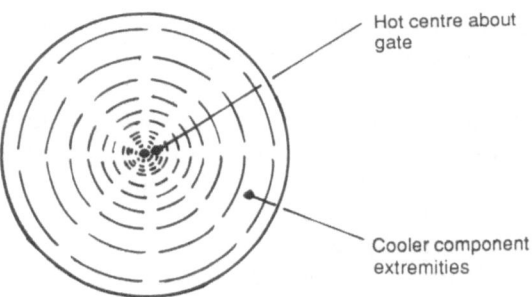

Figure 8.2 Thermal footprint of a plate moulding.

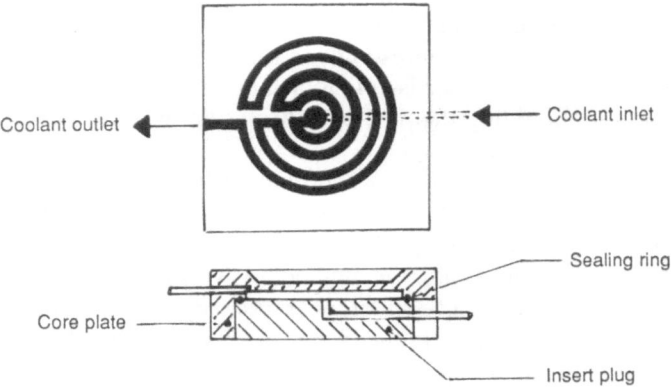

Figure 8.3 Incorporation of spiral cooling circuit.

cooler extremities. To achieve the required thermal balance across the whole moulding a centrally fed spirally circulating cooling channel is incorporated into the mould core design (Figure 8.3).

As the coolant flows through the centrally fed cooling channel, its temperature increases. The progressive temperature increase serves to fulfil the component cooling requirements across the moulding. Additionally the mould cooling profile also serves to aid the filling of the cavity by not over-chilling the polymer at the extremities of the moulding where the cavity pressure is at its least effective.

Due to the complexity of component and mould designs, a cooling balance (as illustrated in Figure 8.3) is rarely as easily achieved. To be able to recognise the thermal footprint of a moulded component the mould designer should consider the following points:

(a) The relevance of gate position and the resultant filling profile. Flow constrictions or other features, such as 90° bends, tend to gain heat due to the shearing effects of the polymer flowing through and about them within the immediate proximity of the gate entry point.

(b) Thick wall sections within the component design, e.g. points where webs intersect, bosses protrude or sections meet, must be considered possible locations where hot spots are most likely to appear.

(c) The position of the ejection system in relation to the component cooling requirement will always result in a contradiction of requirements. Features which include bosses and webs may require individually ejecting to avoid component sticking and damage resulting during the ejection cycle. The inclusion of ejector pins and sleeves on or about such features reduces the space available to position the cooling flow channels which as a result, in some cases, become omitted from the mould design within the feature locality.

(d) The inclusion of inserted features within the mould core and cavity components causes problems (chapter 6). Inserted features such as intricate core and cavity or shut-off details will easily gain heat during moulding. The inclusion of conventional cooling channels in such features, because of their size and location, is often realistically impractical. In such cases other cooling methods have to be employed, e.g. 'cool pins' (thermally conductive rods) – see next section.

8.3 Mould cooling time

The length of cooling time required to solidify the moulding sufficiently can be controlled and reduced if the relevant information is obtained or estimated prior to the designing and incorporation of the cooling system, i.e.

(a) estimated cooling time for the component to be moulded;
(b) the cooling medium to be utilised;
(c) the thermal properties of the polymer to be processed;
(d) the thermal properties of the mould construction material to be used.

All the above factors have a direct influence on the cooling efficiency of the mould and the resultant quality of the moulding produced. The final component production cost is predominantly influenced by the moulding cycle time of which a high proportion is concerned with cooling the moulding (section 8.2).

8.3.1 *Estimation of cooling time*

The cooling time of a component can be estimated (chapter 1). The cooling time estimation is open to error, due primarily to the fact that the formulae and graphs used assume the following:

(a) Total contact exists between the moulding and the mould walls at all times for the duration of the cooling time. With respect to the injection moulding of thermoplastics this is rarely the case, as a result of the shrinkage characteristics of the materials while cooling.
(b) The flow rate and temperature of the cooling medium (or simply coolant) is constant throughout the whole mould resulting in a constantly even mould temperature. In reality the surface temperature of the mould varies according to the moulding conditions encountered and the complexity of the core and cavity build-up.
(c) The moulded component is homogenous throughout its entire cross-section. This is not the case where injection moulded components are concerned, as a result of:
  (i)   the skinning effect of the polymer, as a result of contact with the cooler surface of the mould;
  (ii)   the orientation of the polymer on filling the mould;
  (iii) the addition and orientation of fillers within the polymer, e.g. talc;
  (iv)   the addition and orientation of reinforcements within the polymer, e.g. glass fibres, which tend to congregate at certain points within the moulding;
  (v)   the presence of varying degrees of crystallinity within the moulding;
  (vi)   the presence of voids or air entrapments within the moulding;
  (vii) thermal losses at the ends of the component.

To estimate the cooling time of a component the relevant formulae and cooling curve graphs have to be employed (chapter 2).

Table 8.1 Commonly encountered coolants

Cooling medium	Thermal working range °C
Antifreeze (e.g. water/glycol)	−20–0
Inhibited chilled/heated water	0–90
Heated oil	90–200
Electrically heated	150–450
(usually in conjunction with water or oil, as above)	

8.4 Cooling media (or coolants)

The mould temperature required is dictated by the thermal requirements of the polymer to be processed. To achieve the required mould temperature a range of cooling or heating media are employed. The most commonly encountered are shown in Table 8.1.

Using the correct cooling medium is essential if the required heat removal rate is to be achieved. Maintaining the coolant temperature may prove a costly process in terms of the need for additional ancillary equipment of the required thermal capacity, e.g. individual mould oil heater and water chiller units.

The utilisation of underrated mould temperature control ancillary equipment will detract from the quality output of the mould, especially when long production runs are to be undertaken.

8.5 Conductive thermal properties of mould construction materials

The thermal conductance properties of the mould materials have a direct influence on the thermal efficiency of the mould. This point can be overlooked by the mould designer during the material selection process. Mistakes occur when mould materials of relatively low thermal conductivity are used in conjunction with polymers of relatively high specific heat. Examples of thermal conductance values are illustrated in Table 8.2.

Table 8.2 Thermal conductance of mould making materials

Mould construction material	Thermal conductivity (W/mK)
Ni/Cr mould steels	30–60
Stainless steels (12–18% Cr)	18–13
Aluminium	197
Beryllium, copper	158
Brass	94

With reference to Table 8.2 it is worth noting that the addition of alloying elements to steels has a marked effect on the conductive efficiency of the material for mould making purposes.

8.6   Cooling – design options

After selecting the cooling medium and identifying the most likely hot spot positions on the moulding, the layout of the cooling system can be planned. For large components or components of high accuracy content the use of CAD based packages, e.g. MoldFlow, provides the necessary means with which to analyse the efficiency of the proposed cooling system. The use of such computer based packages can prove extremely effective for pin-pointing both hot and cold spot locations within a moulding.

8.6.1   *Location of cooling channels*

To obtain the maximum effect from the mould cooling system, cooling channels must be accurately positioned and dimensionally correct. The cooling effect about each channel tends to be radial in form (Figure 8.4), drawing heat from the surrounding metal into the channel.

Even thermal dispersion relies on the interlocking of the individual

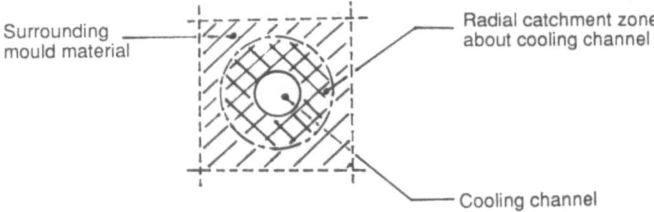

Figure 8.4 Radial cooling effect of a cooling channel.

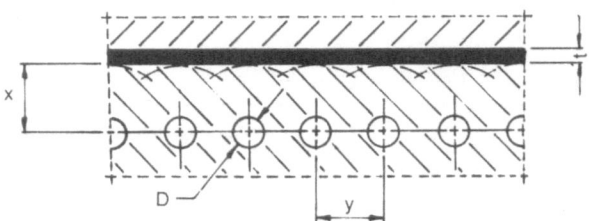

Figure 8.5 Location of cooling channels: $x = 1.5D$, $y = 3D$. For component wall thickness: $t = 2\,\text{mm}$, $D = 8\,\text{mm}$; $t = 3\,\text{mm}$, $D = 10\,\text{mm}$; $t = 4\,\text{mm}$, $D = 12\,\text{mm}$.

catchment zones about the channels. Due to the symmetry of the radial cooling effect, a simple formula can be applied to aid the laying out of cooling channels (Figure 8.5).

8.6.2 *Flat plane cooling*

Plane or plateau cooling relates to a cooling system in which the cooling channels lay next to each other within the same level or plane (Figure 8.6). This design of system can be employed to cool large flat mouldings of uniform cross-section, e.g. a plastic tray moulding. Two distinct feeding methods are employed, one connected in parallel, the other in series.

8.6.3 *Spiral cooling*

Spiral cooling systems are frequently employed for the cooling of centrally gated mouldings, e.g. plates, bowls and buckets. The coolant enters the system at the centre (the hottest point) then circulates to the extremities of the moulding to exit (Figure 8.7).

For smaller centrally gated components such as flower pots and pack-

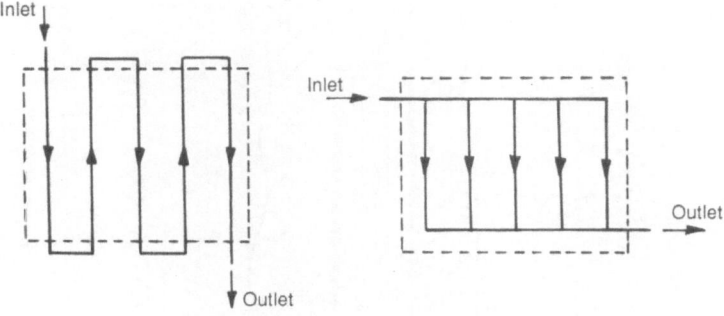

Figure 8.6 Series and parallel plane cooling.

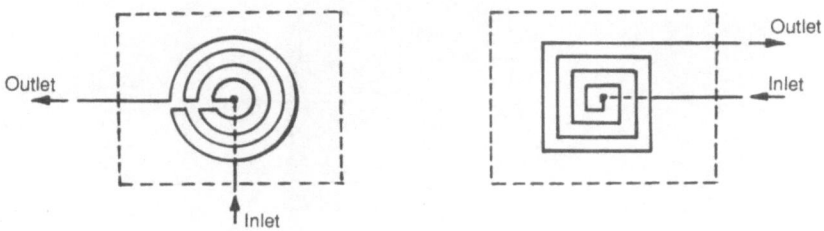

Figure 8.7 Round and boxed spiral cooling.

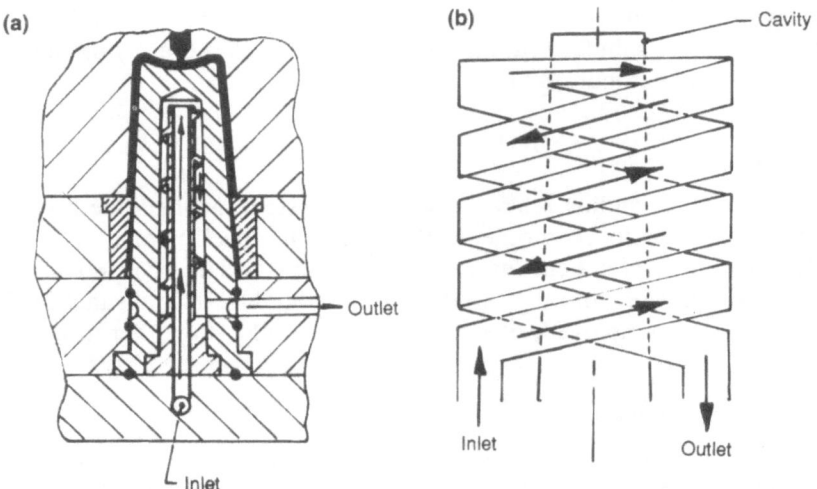

Figure 8.8 Spiral core and cavity inserts. (a) Core – centrally fed spiral cooling channel. (b) Cavity – externally fed reverse helix jacket insert.

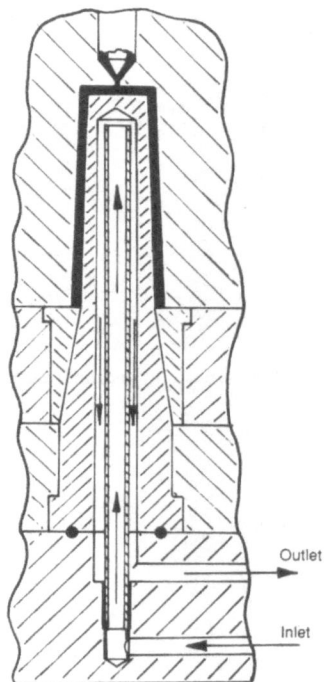

Figure 8.9 Finger cooling of a narrow mould core.

aging, spiral core and cavity inserts can be used (Figure 8.8). The spiral flow channel is externally machined into the core insert sleeve, a parallel 'O' seal seating groove being machined into the end of the insert to create a leakproof seal once fitted into the core body.

8.6.4 *Finger or bubbler cooling*

Long thin core or cavity sections of reduced thickness, e.g. tube mouldings, are often too thin to accommodate spiral inserts or the like. The centre of the core feature is drilled out and a hollow brass tube held in the centre of the orifice (Figure 8.9). Coolant is forced up the centre of the tube to exit at the top of the core and the coolant backflows outside the tube walls to exit at the base of the cored feature. This system is ideally suited to components gated to the opposite face at the top of the core, thus providing the most effective heat removal from the moulding directly opposite the gate.

8.6.5 *Baffle cooling*

Baffled cooling channels are widely employed to cool narrow sections, e.g. between webs, or larger areas when used in numbers, e.g. around a

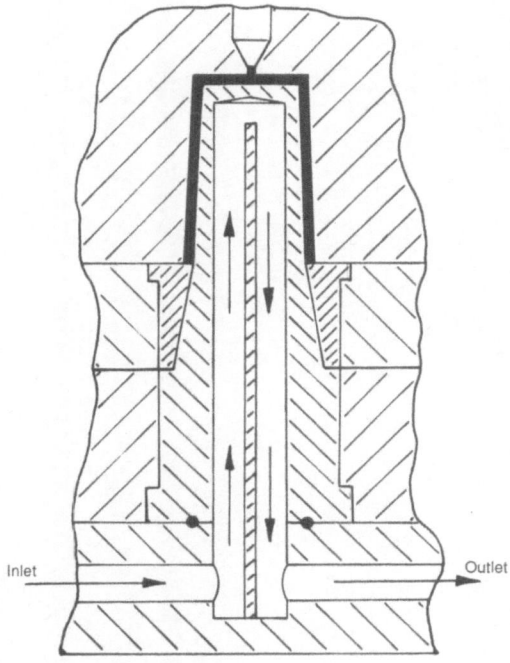

Inlet   Outlet

Figure 8.10 Baffle cooling of a tube moulding.

cavity. A blind hole is drilled at 90° through a cross-drilled cooling channel. A machined brass plate is inserted into the centre of the hole, dividing it into two equal sections. Coolant is pumped into one side of the baffle, which deflects the flow up and over the feature at the top of the blind hole, exiting the other side at the baffle base (Figure 8.10). Baffles, like bubblers, can be fed either in series or in parallel, as required.

The general construction of a mould cooling system can include many designs of individual cooling circuits. Flow control of individual circuits is essential for certain mould features, e.g. cavity bases, side walls and cores, dependent upon the quality requirements of the moulding.

9 Ejection

9.1 Introduction

The function of an ejection system is to enable the removal of the moulded component from the mould once solidified. Ejection mechanisms vary enormously, both in function and design. Choice of ejection method is usually dependent on a number of factors, for example:

(a) design of component to be ejected;
(b) aesthetic considerations of the component;
(c) production requirements.

It will be evident that careful consideration must be given to the design and choice of ejection method when designing both components and moulds for injection moulding purposes.

9.2 Choice of ejection method

Before selecting an ejection method, sufficient thought must be given to the various component and production requirements. It may prove beneficial to construct initially a short list of requirements with respect to the component to be moulded. Such a list should include references to the following:

(a) Component aesthetics.
  (i) Can ejector pin witness marks be tolerated on visual presentation faces? Reversing the impression within the mould design is not an uncommon practice to overcome this problem.
  (ii) Does sufficient draft angle exist on the mould cavity walls to avoid component 'scuffing' or 'dragging' occurring during ejection? Adequate drafting of any moulded component is essential and should be considered as part of the component design prior to the designing of the mould.
(b) Component dimensional considerations.
  (i) What amount of ejection-induced distortion can be tolerated dimensionally? Moulded components can easily be distorted during ejection if unsupported and especially in the case of thin walled mouldings.

(ii) Will additional support webs or ribbing be required to enable ejection to take place without distortion resulting?

(c) Machine considerations.

(i) Is the ejection force sufficient to eject the component within the working limitations of the moulding machine? Mould designers will often design a whole mould without giving ejection force requirement a thought, resulting in disastrous consequences.

(ii) Is the intended moulding machine able to actuate the designed ejection system incorporated into the mould design? Machine designs may vary with respect to ejection actuation method. The majority of moulding machines ejection actuate on the centre line of the platens in line with the injection unit (chapter 4). Whereas a minority of machines actuate by means of cross head plungers or puller plates, very large machines tend to employ cross-headed assemblies to distribute the larger forces as evenly as possible.

(iii) Will additional 'core pulling' or actuation circuits be required? With complicated mould designs, multi-core pulling may be a design requirement (chapter 12). For this purpose the moulding machine should possess additional hydraulic core pulling circuits and the necessary means of actuation control.

(iv) With the moulding machine/mould combination fully open, will sufficient 'daylight' (open space) exist to allow for component ejection and clearance? Failure to allow for sufficient daylight will mean that the mould will have to operate in a larger, more costly machine with a greater opening stroke. The running of a smaller mould in a large machine could lead to the following problems:
   - material degradation, caused by longer barrel residual times;
   - higher levels of mould wear as a result of increased clamping pressures;
   - component flashing, created by the 'rocking' of the platens on closure with the smaller projected area of the mould;
   - shot inconsistency due to the utilisation of a large volume injection unit delivering a small shot volume;
   - increased cycle times; larger machines tend to move more slowly than their smaller counterparts.

(v) Is the required ejection stroke within the working capability of the moulding machine? Mouldings of greater draw depth, e.g. tubes, etc., will require long ejection strokes with which to eject them. The maximum available ejection stroke on the moulding machine is determined by the length of the ejector cylinder (i.e. the swept volume of the piston on actuation).

This has to be investigated by the mould designer if stroke length is in doubt.

(d) Mould/tooling considerations

    (i)   Is the ejection system to be employed additionally for cavity venting purposes? Ejector pins, etc. are frequently added to a mould design to vent gas traps where blind spots occur within the cavity.

    (ii)   Will the ejection system be required to release interfacial vacuums between the component and the mould core prior to the component being ejected? Large thin-walled mouldings, such as lids, fully encase the core on mould opening and a vacuum is created underneath the component during ejection. Thin-walled brittle mouldings are easily cracked or broken by the holding force of the vacuum if not released or vented in time.

    (iii)   Ejection components such as pins and blades induce high levels of wear to the mould core components during service.

9.3 Ejection methods

9.3.1 *Pins and blades*

Ejector pins provide one of the cheapest forms of ejection available although their use has limitations. One such limitation is that due to the very small projected area of the pins, high single point loadings are

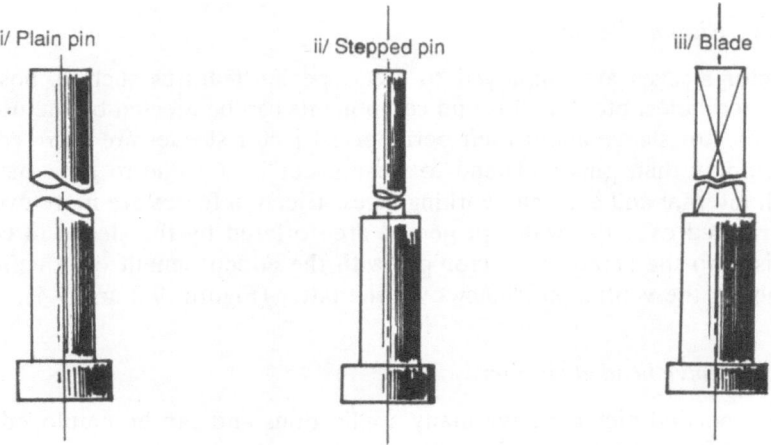

Figure 9.1 Ejector pin types.

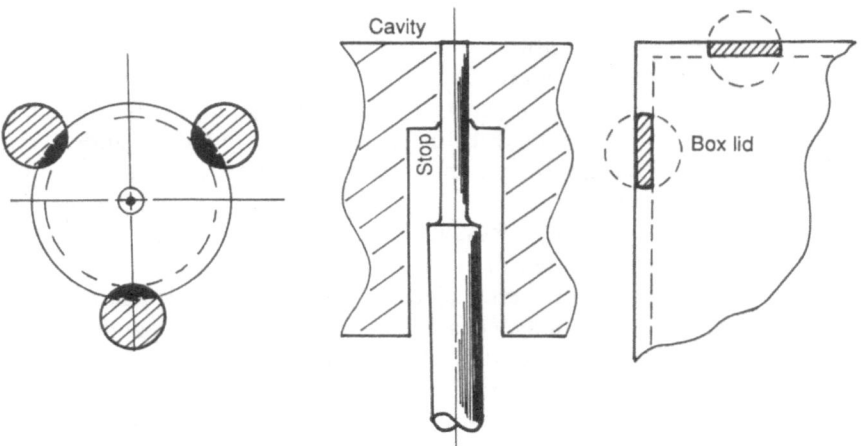

Figure 9.2 Pin and blade positioning.

transferred to the moulding during the ejection cycle. Component damage or distortion may result where pins are located, especially in the case of thin-walled mouldings or when brittle materials are used.

Ejector pins are usually fitted as standard mould components due to their cheapness and ease of availability from standard mould part suppliers. Pins are manufactured to standard formats in metric and imperial sizes; the standard formats for blades, plain and stepped pins is illustrated in Figures 9.1 and 9.2.

9.3.2 *Ejector sleeves*

Ejector sleeves are employed to eject specific features such as bosses, recessed holes, etc. Small round components can be ejected by the use of one ejector sleeve about their periphery. Ejector sleeves are more costly to employ than pins and tend to wear faster in use due to their having both internal and external working faces. Ejection forces are more evenly distributed over the wider projected area offered by the sleeve in comparison to the standard ejector pin with the added benefit of effectively doubling the venting efficiency over the latter (Figures 9.3 and 9.4).

9.3.3 *Valve headed ejectors*

Valve headed ejectors have many applications and can be employed for many different reasons. The various designs of valve ejector are usually concerned with the fixing or actuation method employed. Valve ejectors offer large projected areas for load transmission. The flat head design

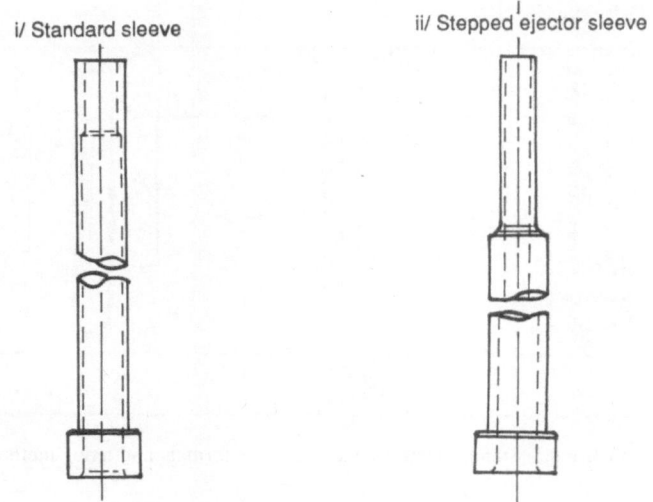

Figure 9.3 Standard sleeve formats.

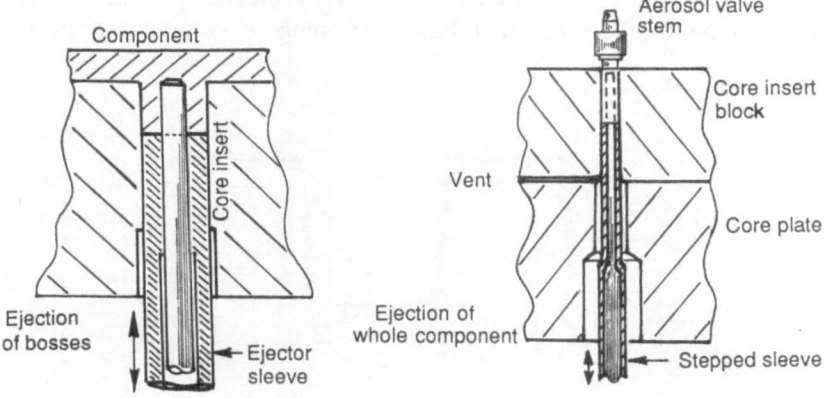

Figure 9.4 Sleeve positioning.

ensures that adequate support can be provided to thin-walled components without undue distortion resulting on actuation. Soft flexible polymers and components, e.g. HDPE bottle caps and closures, frequently employ this form of ejector to form seal undercuts and aid ejection. Floating valve ejectors can be employed in vertically used injection moulds or compression moulds utilising gravity to return them after actuation (Figure 9.5).

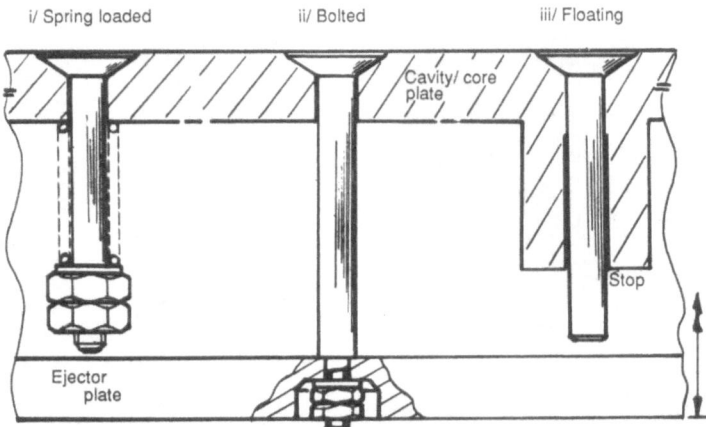

Figure 9.5 Standard valve headed ejector formats and fixing method.

9.3.4 *Stripper ring and plate ejection*

Stripper rings and plates (Figure 9.6) eject components by pushing or pulling them off the mould core, usually by means of total peripheral contact about the component base. Stripping of components provides

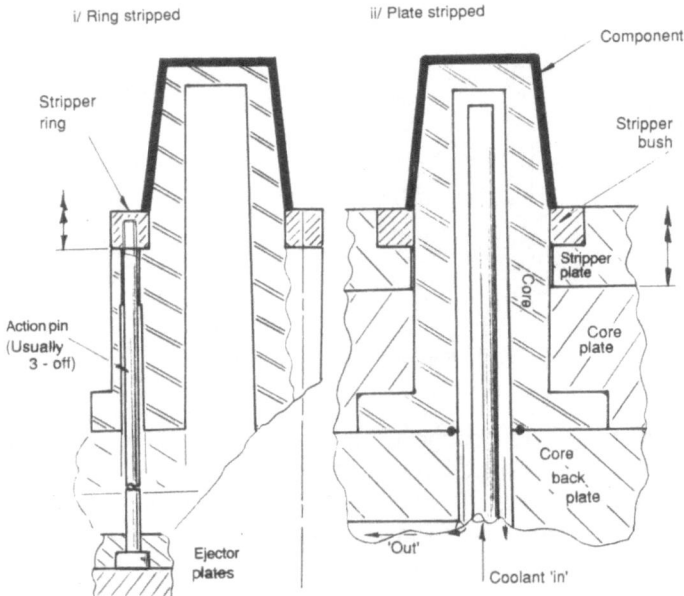

Figure 9.6 Stripper ring and plate formats.

benefits in terms of good component support during the whole ejection phase resulting in less distortion or damage being inflicted. The use of such ejection methods, although beneficial in many ways, can result in higher levels of tool wear occurring due to the increased size of the ring or plate bush locational features about the mould cores.

9.3.5 *Air ejection*

Combinations of the various mechanical ejection methods are often coupled with the use of air (pneumatic) ejection when the need arises. Many examples of the use of air/mechanical ejection (Figure 9.7) can be found within the thermoplastic and rubber industries for components such as gaskets, seals, diaphragms, etc. Large, thin-walled lightweight components often utilise air ejection techniques due to their large projected areas, which make conventional ejection difficult if component damage is to be avoided.

9.4 Estimation of ejection force

The amount of force required to push a moulding off a projecting core or out of a cavity can be estimated. Most mould designers will suitably

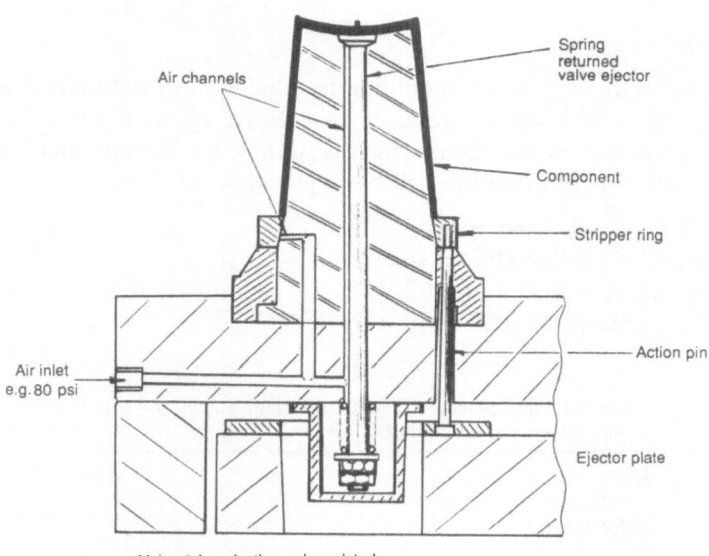

Valve / ring ejection - air assisted

Figure 9.7 Mixed ejection methods with air assist.

design an ejection system and assume that it is fit for the purpose. The machine specifies how much ejection force is available. Mould manufacturers have learnt to their cost (especially when dealing with deeply drawn components) that quite high ejection forces can be encountered.

A simple formula can be used to estimate the ejection force required:

$$P = \frac{\delta t \times E \times A \times m}{d\left(\dfrac{d}{2t} - \dfrac{d}{4t}n\right)}$$

where

P = ejection force (kN)
E = elastic modulus of polymer (N/cm^2)
A = total surface area in contact with moulding and mould face in the line of draw (cm)
m = coefficient of friction, plastic on steel
d = diameter of circle of circumference equal to the perimeter of the moulding (cm)
t = thickness of the moulding (cm)
n = Poisson's ratio
δt = thermal relationship of plastic across diameter of projection or cavity. (δt = coefficient of expansion of polymer $\times d \times \Delta T$) where ΔT = plastic softening temp. − mould temp.)

Example

A single impression cover moulding is required to be moulded in polystyrene on a 250 tonne press. If the stated ejection force is 30 kN, estimate whether or not the machine is suitable for the intended purpose.

The following information will be required:

Material to be moulded = PS
Coefficient of friction (PS on steel) = 0.4
Poisson's ratio = 0.35
Modulus of elasticity (PS) = 30 × 10^4 N/cm

Table 9.1 Approximate friction coefficient values for some commonly encountered plastics on steel

ABS	0.24
Acetal	0.13
Nylon 6,6	0.2
PC	0.25
PVC	0.22
SAN	0.29

Coefficient of expansion for PS = 7×10^{-5} cm/°C

Softening temp. of PS = 80°C

Mould temp. for PS = 20°C

Component dimensions = rectangular box 12 cm × 23 cm × 9.5 cm, deep wall thickness 0.3 cm STD

Area for line of draw = sum of the sides

$A = (12 \times 9.5)2 + (23 \times 9.5)2 = 665\,cm^2$.

$$d \text{ (box perimeter)} = \frac{(12 \times 2) + (23 \times 2)}{\pi} = \frac{70}{\pi}$$

Since $d\pi = 70$

Then $d = 70 \div \pi = 22.3$ cm.

δt (thermal relationship) = coeff. of expansion (PS) $\times D \times \Delta T$

$$\therefore \delta t = 7 \times 10 \times 22.3 \times (80 - 20)$$
$$= 0.094.$$

$$\text{Ejection force} = \frac{0.094 \times (30 \times 10) \times 665 \times 0.4}{22.3\left(\dfrac{22.3}{0.6} - \dfrac{22.3}{1.2} \times 0.35\right)} = 11\,KN$$

The estimated answer is well within the working capability of the press.

10 The three-plate mould

10.1 Introduction

The three-plate mould differs from the more common two-plate design format in terms of utilising more than one split or parting line. The tool construction is divided into three distinct plate build-ups which separate from each other on opening (Figure 10.1). One opening provides clearance for component ejection, while the other allows for sprue ejection and clearance. Being a tool of increased complexity, the three-plate mould is therefore more time-consuming and expensive to manufacture than its two-plate cousin.

10.2 Why choose a three-plate mould?

The main reason for choosing the three-plate tool layout is the flexibility which the design offers in terms of gate location. The three-plate configuration enables the inclusion of multi-gate positions on larger mouldings or the centre gating of smaller components to produce better quality mouldings. The more recent emergence of runnerless or hot runner mould (chapter 11) designs has largely reduced the practicality of adopting the three-plate design format for most moulding operations. By comparison with the various hot runner mould designs, the three-plate design still offers a few advantages, usually in terms of the following:

(a) *ease of material or colour changing during use*; the three-plate mould clears its feed system every working cycle giving fast colour changes without the problems of long-term material contamination.
(b) *reliability*; the relative complexity of many hot-runner tool designs, especially their heating and control systems, makes them prone to electrical failure and subsequent breakdown. Three-plate moulds, having very few electronic components, tend to be more reliable once set and running although having more mechanical moving parts, i.e. linkages, bearings, etc., the three-plate design tends to be more prone to mechanical failure if not correctly serviced.
(c) *cheaper initial capital outlay*; mould heaters and temperature control equipment can be very expensive, often resulting in high initial capital expenditure. For shorter production run requirements, the

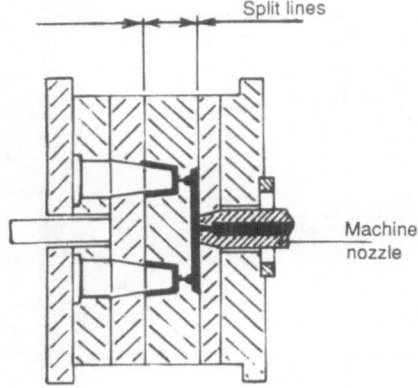

Figure 10.1 Schematic diagram of three-plate mould.

additional capital outlay may not be financially viable and the three-plate option could provide a cost-effective alternative.

(d) *the moulding of thermally sensitive polymers*; the thermal sensitivity of some polymers could dictate the need to process the material on a conventional design of mould tool, i.e. non hot runner. In such cases two-plate mould designs are conventionally employed but in instances when the component gate location demands off-the-edge siting the three-plate conventional format is worth consideration.

Although the three-plate mould design does offer advantages under certain circumstances, generally the design tends to be inferior when compared to the production efficiency of the various hot runner designs available (chapter 11).

10.3 Tool construction

Figure 10.2 shows a typical buildup of a three-plate mould in which each individual impression is filled through a centrally situated pin point gate. The mould can visually be divided into three sections:

A – the feed plate assembly;
B – the intermediate or cavity plate;
C – the core plate assembly.

The general construction of the tool is built about the opening sequence of the mould. To fully understand the principles involved in designing such a mould, we must firstly understand the design requirements of the mould opening sequence. The opening of the mould is usually undertaken in three stages:

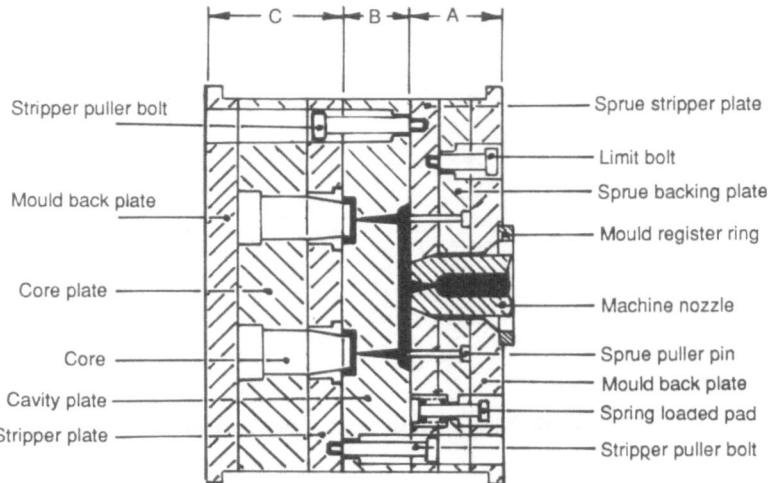

Figure 10.2 Construction layout of a three-plate model.

Stage 1   As the machine platens open, the intermediate plate (B) is pulled away from the feed plate assembly (A). The runner system is retained on the feed plate assembly by runner puller or sucker pins. The retaining of the runner on the feed plate section serves to break the gate away from the solidified mouldings while being fully supported by both core and cavity components. This stage is known as the *breaking* first stage or sequence.

Stage 2   With the gates parted from the mouldings and the runner clearance daylight fully extended, the mould is ready for the next stage which is runner stripping and ejection. With the runner daylight fully extended, the runner stripper plate is pulled forward by the engagement of the intermediate plate shoulder limit bolts. The runner system is then stripped off the puller pins as the runner stripper plate is pulled forward. The length of runner stripping stroke is determined by the engagement height of the runner plate shoulder bolts. This stage of the tool opening sequence is known as *runner stripping*.

Stage 3   The ejection of the mouldings can be started by the opening of the core and cavity split line once movement of the intermediate and feed plate assemblies has been arrested. With the moulding daylight sufficiently extended the mouldings can either be ejected conventionally (chapter 9) or stripped off the cores by the forward pulling action of the core stripper plate as the mould reaches its maximum opening stroke.

Note: stages 2 and 3 are often synchronised to function together, dependent upon the design of tool used.

The explanation of the three-plate mould opening sequence serves to highlight one of the drawbacks encountered with this design of tool, namely long opening stroke requirement. The major limiting factor concerned with employing any design of multi-daylight tool has to be the maximum opening stroke available on the intended moulding machine. For this reason it is wise to calculate the required working daylight clearances before embarking upon the chosen course of action, especially when dealing with deeply drawn mouldings, e.g. test tubes or the like.

In order to gain control over the required mould opening sequence, platen position and speed of movement has to be accurately set and controlled throughout the moulding cycle and production run. In general, the closing of the three-plate mould is achieved by the closing action of the moulding machine platens which progressively pick up and close the mould sections as they move forward.

11 Runnerless moulds

11.1 Introduction

A runnerless or hot-runner mould system could be considered as an extension of the moulding machine injection unit. In this case, the polymer is kept molten right up to the cavity gate by means of additional heating elements controlled by thermocouples. The working temperature of the hot-runner system is not dissimilar to that of the molten polymer being processed. In order to maintain a stable running temperature, the hot feed system is either insulated from the rest of the mould, e.g. by an air gap, or controlled by the addition of cooling channels within the immediate vicinity. The choice of method depends on the design of hot feed system adopted.

11.2 Advantages of the hot-runner mould

The principal advantages of employing a hot-runner mould are:

(a) reduced cycle times as a result of having a component cooling requirement only. The runner and feed system remain molten above the quickly frozen gate.
(b) material savings result from having no sprue or runner systems to granulate or dispose of. Without the subsequent regrind addition to the initially fed material, shot and production consistency is maintained throughout the production run.
(c) labour and post-moulding finishing costs are significantly reduced without the need for degation of the mouldings.
(d) the ability to gain greater control over the mould filling and flow characteristics of the molten polymer during the filling phase of the moulding cycle. This can be achieved by locally altering the temperature of the feed system within the vicinity of the subject area.

In the case of the hot-runner mould, the relative advantages usually outweigh its limitations. Certain limitations such as colour changing and reliability have to be carefully considered when deciding on either a conventional mould or the hot-runner alternative. In cases where high volumes of mouldings are required at low cost, the hot-runner option often provides the only logical route to take.

11.3 Hot-runner systems

Many designs of hot-runner mould exist. They tend to vary according to system design and individual manufacturers. Rarely are any two manufacturers' systems compatible with each other, either in size or design of assembly. Variations exist in terms of the method of heater supply and control, some being AC supplied, others being DC supplied, and requiring their own special heater/controllers and built-in transformer units. After surveying all the hot-runner mould types and variations, there appear to be three general groups into which they may be categorised.

Group 1 – Externally heated manifold moulds.
Group 2 – Internally heated manifold moulds.
Group 3 – Insulated hot-runner moulds.

In addition to the above groupings, individual mould designers and system manufacturers frequently utilise design features from more than one group. For this reason group 4 has been added.

Group 4 – Mixtures of the aforementioned.

11.4 The externally heated hot manifold mould

Figure 11.1 illustrates a typically encountered construction layout of an externally heated manifold hot-runner mould. The design centres about the 'hot' manifold block (HMB) into which a cross-drilled runner/feed system is machined. Each runner flow channel is drilled and plugged off

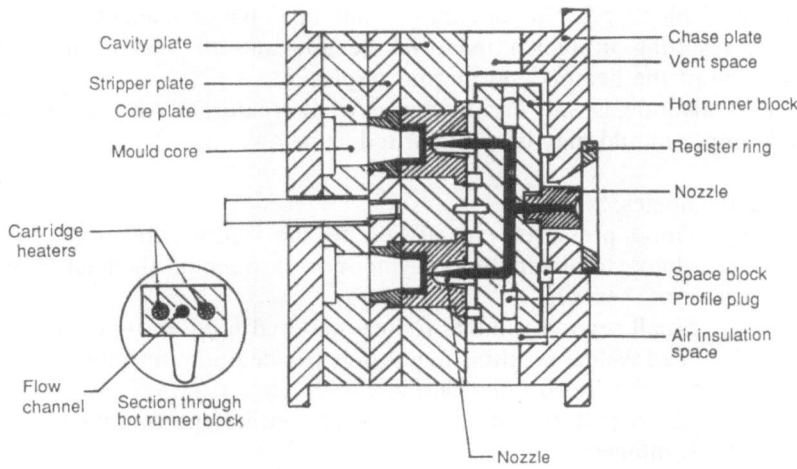

Figure 11.1 Typical externally heated hot manifold mould.

accordingly until the required runner layout is achieved. Care must be taken when designing the runner layout of the manifold block to ensure that flow 'dead' or 'blind' spots are not created where degradation of the polymer could occur in use. Features such as profiled end plugs can be incorporated into the right angled corners of runner intersections to reduce the chance of flow dead spots being created. Heat is applied to the manifold block by means of cartridge or bar heaters sited in drilled holes about the melt flow channels. Correct positioning of the heater sites is essential if even thermal saturation of the block is to be achieved. Thermal control of the block is aided by the location of thermocouples sited in between the heating elements and the flow channels. These are in addition to those already built into the heaters themselves. In order to maintain a thermal balance between the manifold block and the rest of the mould, an air insulation gap should be incorporated into the mould design. A gap of approximately 6 mm is normally considered sufficient for general moulding temperatures. The gap is achieved by sealing washers about the flow channel outlets at the front of the block and support blocks situated at the rear of the block against the mould back plate. The manifold is usually centred to the rest of the mould by dowel pins located into the front of the block which allow expansive forward movement to occur, but resist lateral movement of the block across the sealing faces of the compression washers. In order to achieve a good sealing shut-off between the manifold sealing washers and the backs of the cavity gates, a shut-off contact interference loading is required. The interference loading is generated by the thermal expansion of the manifold block and the BeCu sealing washers which crush under load to form a compression seal against the backs of the gates. Failure to sufficiently 'heat soak' the manifold block prior to use may result in polymer leakage at the seal faces resulting in blown manifold heaters and the need for a costly rewiring of the heaters and thermocouples.

The advantages and disadvantages of the externally heated manifold hot-runner mould may be summarised as below.

1. Advantages:
   (a) Good pressure transmission to the cavity gate as a result of almost totally molten polymer flow throughout the heated runner cross-section.
   (b) Small section runners reduce material hold-up times within the feed system resulting in less polymer degradation when compared to other hot-runner designs.
   (c) Consistent output once thermally settled and running.
2. Disadvantages:
   (a) Susceptible to material leakages at sealing washer seats resulting in an additional mould servicing requirement, i.e. replacement of the BeCu sealing washers between runs.

(b) Preheating or 'soaking' of the manifold block prior to use: Preheating of the manifold usually takes approximately $\frac{3}{4}$ hour to ensure that a leak free shut-off exists between the block, washers and gates.

(c) Feed control of individual impressions is limited due to the general heating effect of the design. If the feed control needs to be individually adjusted to any singular impression, the gate geometry or bore diameter of the sealing washer is altered to suit.

(d) Less economical to run than other hot-runner designs, in terms of higher electrical heating power input required, in comparison with the processing output of the design.

11.5   The internally heated manifold mould

In the case of the internally heated hot-runner mould, molten polymer is routed about the hot-runner or distributor block (Figure 11.2) contained within large diameter (32 mm typical) cross-bored flow ways. Steel tubes containing cartridge heaters with built-in thermocouples are centrally supported by end caps secured in position at the end of each flow bore and usually anchored by either dowel pins or shoulder bolts. On entry into the mould, the molten polymer flows about the heated tubes as it flows through the feed system, remaining molten while in contact with the heated tubes and freezing to form a skin within the cooler channel outer walls. Skin thickness is determined by the temperature of the surrounding hot-runner block (HRB) processing temperature of the molten resin and the hold-up time experienced during cycling. To achieve a working thermal balance, the temperature of the hot-runner block must be controllable

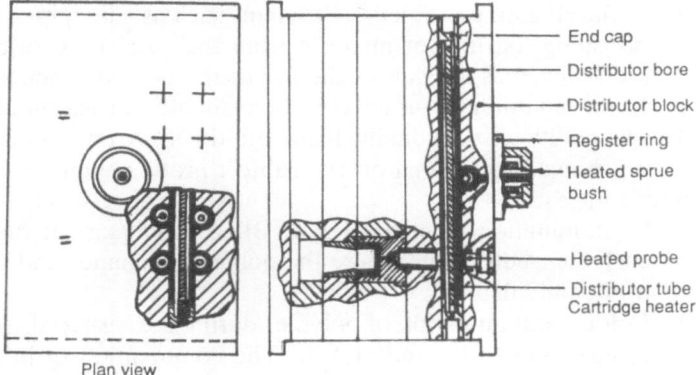

Plan view

Figure 11.2 Typical internally heated manifold mould.

about the distribution tubes. For this purpose drilled cooling channels are incorporated into the block design. Positioning of the block cooling channels should be such as to remove excess heat from behind the block to the moulding machine platen and to the front of the block, to the mould cavity plate. Failure to dissipate excess heat from the HRB effectively reduces the size of the processing window available during use. Quality output from the mould will also suffer accordingly.

Filling control of individual cavities is achieved by the incorporation of a heated tipped probe behind each gate. Each probe is individually heated by its own cartridge heater and controlled by an individual thermocouple. Probe height is determined by altering the thickness of the probe seating washers located under each probe shoulder. By altering the probe height and the heater temperature, the filling of the cavity can be accurately controlled. Even relatively unbalanced feeding layouts can be set to fill evenly using such a system.

The effective height of the probe in use has to be calculated prior to setting the mould up and account has to be made for the thermal expansion of the probe material (usually alloy steel) when at running temperature. When in use, the probe is subject to natural wear and tip damage and should therefore be considered as a standard mould part with spares held accordingly. In order to reduce setting and servicing costs, individual probe setting heights and running temperatures should be noted and recorded for future reference and information purposes.

The advantages and disadvantages of the internally heated manifold mould may be summarised as follows.

1. Advantages:
    (a) Good filling control to individual cavity gates. The use of individually heated and controlled probes behind each gate enables accurate and balanced filling control to occur across the whole mould once set.
    (b) Reduced gating problems. Problems such as gate 'pips', material 'whisping' (strings of material from the gate) or gate blockage are reduced as a result of the use of the heated probe set-up.
    (c) Good moulding accuracy control available to each moulding.
    (d) The ability to individually 'blank off' defective impressions which is achieved by turning off the subject probe heater.
2. Disadvantages:
    (a) Contamination from the HRB. Blind spots exist at the end of each distributor tube where the polymer stagnates and degrades if not solidified.
    (b) Colour contamination of polymer during use, especially if colour changes are to be undertaken. The combination of blind spots and the skinning effect of the polymer inside the flow bores are

all sources of possible colour contamination which are more pronounced if higher processing temperatures are encountered later on during mould use.

(c)   Probe tip wear and damage during use. Probe tips are easily damaged by the gate walls if the probe is inadequately supported when in use, most of the damage occurring during the higher pressure first stage of the moulding cycle. The tips naturally wear in use due to the abrasive action of the pressurised polymer, especially if reinforced or filled in any manner.

(d)   Gate blockage. Degraded polymer or foreign matter usually collects between the probe tip and the cavity gate wall which eventually results in a blocked gate.

11.6   The insulated hot-runner mould

The insulated hot-runner mould (Figure 11.3) is the simplest of all the hot-runner designs. The design relies upon the principle that polymers are good insulative materials with relatively high specific heat capacities. Molten polymer is initially injected into the flow channels and a skin is formed on the polymer in contact with the cooler channel walls. The skin serves to effectively insulate the molten core of the feed material as it progresses through the mould to the cavity gates. Flow channel skin thickness is determined by the temperature of the mould plates, temperature of the polymer flowing through the system and the length of the moulding cycle time. With this tool design the mould feed block is split onto two plates allowing access to the feed channels once taken apart. The large diameter (25–35 mm typical) feed channels are half form machined into each plate using a bull-nosed end milling cutter, the full diameter being created on assembly of the two plates. Accurate thermal

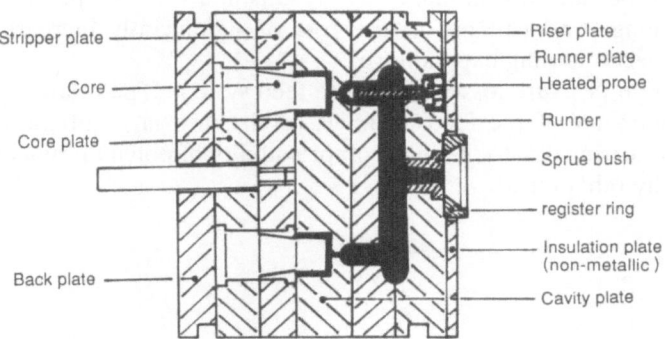

Figure 11.3 Typical insulated hot-runner mould.

control of the feed plates is essential if the mould is to function consistently while in use; coolant flow channels must be provided in front of and behind the feed channels to obtain control. Insulation boarding, i.e. fibre glass plate (typically 6–10 mm thick), should also be added to the mould backing plates to minimise thermal loss to the machine platens.

The advantages and disadvantages of the insulated hot-runner mould may be summarised as below.

1. Advantages:
    (a) The feed system can easily be stripped and cleaned, resulting in very little material or colour contamination occurring. Feed system cleaning or 'deslugging' is often carried out in the moulding machine. Access to the solidified feed system is obtained by unbolting and latching the frontal plate to the moving half of the mould and opening the press.
    (b) Mould start-up times are faster when compared to other hot runner mould systems.
    (c) Moulds are very much cheaper to manufacture than the other hot runner mould designs.
    (d) Thermally unstable polymers may be processed using such a system.
2. Disadvantages:
    (a) Freezing-off of the feed system. Stoppages during the production run will result in the solidification of the feed system and the need for its removal before production can restart or continue.
    (b) Gate blockages may be caused by a frozen 'plug' of material caught in the cavity gate left from the previous shot. Gate 'plugging' is a serious problem with the insulated hot-runner mould design. To overcome this problem, heated probes are often situated behind each gate (section 11.5). The addition of probes into this mould design results in a very competent mould being produced but at the cost of significantly increasing the overall tooling expenditure.
    (c) High pressure losses within the feed system. The relatively large diameter of the feed channels results in a large pressure drop occurring due to the high compressibility of polymer melts generally (chapter 3).

12 Undercut moulds

12.1 Introduction

In order to understand the various designs of undercut mould, we must first understand the definition of what a moulded undercut really is. In its most simplistic form, an undercut can be described as any feature on the moulding which resists the line of draw or inhibits the ejection of the moulding from the mould. Moulded features which include retention beads, grooves, side holes, deeply eroded surface finishes, reverse draft angles, internal and external rigid threads can all be classified as undercut features if they resist the line of draw (Figure 12.1). In order to eject such features from the mould, the forming component parts of the mould must be retracted prior to ejection taking place. This pre-ejection removal of the mould coring components is usually referred to as core pulling.

12.2 Core pulling

Core pulling is an expensive business and should only be undertaken when no alternative option exists. The need to core-pull a feature has to be identified at a very early stage, before the mould build-up has been started. To enable a cored feature to be pulled out of the line of draw, sufficient space within the mould bolster must exist in which the pulling mechanism can be actuated and housed. The amount of movement required to clear the undercut feature, e.g. length of screw thread to be unwound or the length of the side core pins to be retracted, plus clearance, has to be predetermined. Additionally, the design of the pulling actuation system should also be considered and its movement requirements allowed for when deciding upon the size of the bolster. In order to do this, the mould designer must have a knowledge of the various designs of actuation systems available and make a choice of which one is best suited to the required application.

12.3 Core pulling actuation methods

12.3.1 *Cam pin actuation*

Cam pin actuation is one of the most commonly encountered core pulling methods, being easy to install and relatively cheap to obtain as a set of

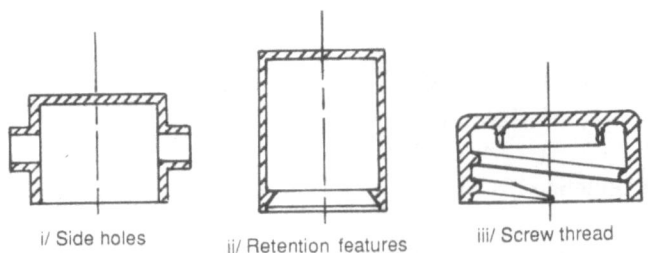

i/ Side holes   ii/ Retention features   iii/ Screw thread

Figure 12.1 Typically encountered undercut moulded features.

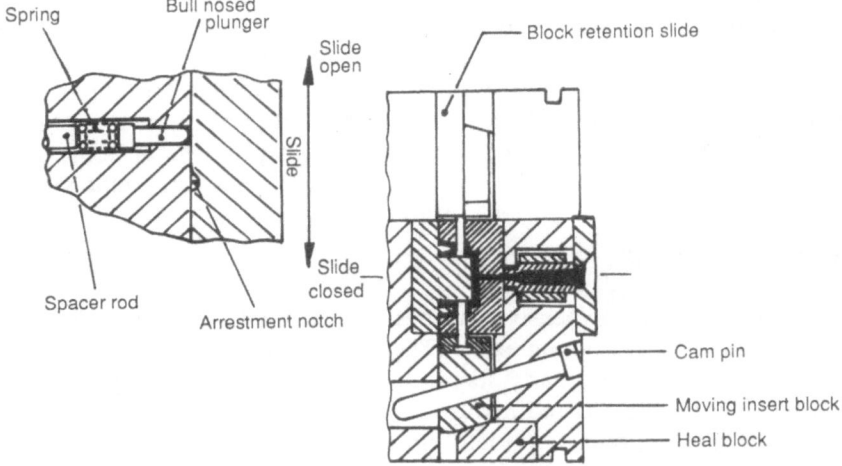

Figure 12.2 Cam pin actuation.

standard mould components (chapter 13). The cam pin assembly (Figure 12.2) consists of four basic component parts:

(i)   the moving insert block;
(ii)   the block retention slides;
(iii) the heal block;
(iv)   the cam pin.

When fitted to the mould, the cam pin and heal block components are usually situated in the cavity half of the tool. The insert block and slide assembly are installed in the core half of the tool facing the cam pin and heal block assembly. On closure of the mould, the angled cam pin locates into the correspondingly angled hole in the sliding insert block. As the mould proceeds to close, the cam pin slides into the angled insert hole forcing the block to move toward the core shut-off. The cam and block assembly is set in such a manner as to enable the side core pin to fully contact the mould core shut-off, without excessive contact force being transmitted. The amount of moving insert block movement is governed by the engagement angle of the cam pin assembly. Angles vary from as low as 10° to as high as 45° to the vertical, dependent on the amount of side movement required. In order to minimise cam pin and block wear, cam pins should be kept short and the engagement angle should be as shallow as possible. Cam action assemblies are also susceptible to damage if the wrong mould closing sequence is used, i.e. the cam pin is misaligned with the insert block hole prior to closure of the mould. In order to minimise the risk of misaligning the cam pin assembly, the following pre-emptive measures can be taken:

(a) Preload the moving insert block to ensure the block is held in the fully open position with the mould open. This is achieved by attaching a spring to the block which is loaded when the mould is partially or fully closed.

(b) Arrest block movement in the open position with the aid of spring loaded ball catches (Figure 12.2) located underneath the moving block. The ball catch loading is only sufficient to suppress minor movement of the block encountered during normal processing conditions. The inclusion of ball catches also serves to reduce wear generated by minor misalignment of the cam pin assembly when in use.

(c) Use a 'mould safety' micro-switched circuit (Figure 12.3), actuated by the positional location of the moving insert block. The movement of the moving insert block serves to open and close the circuit micro-switch. With the block in the fully open position the switch is

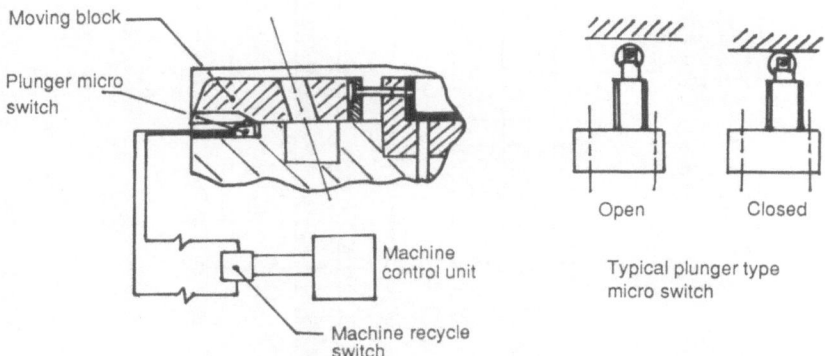

Figure 12.3 Typical side action mould safety circuit.

depressed, the circuit is said to be 'made', and the mould may close without damage resulting. If the block is not in the fully open position and the micro-switch is undepressed or 'not made', an open circuit exists. Since the mould safety circuit is linked in series to the recycle switching circuit of the moulding machine, the machine will not receive a recycle signal and will stop with the platens in the open position. The machine will then sound the alarm signal to inform the operator that something is amiss. The insert block can be accordingly reset to allow production to continue.

12.3.2   *Lost action cam pins*

These are often referred to as 'dog leg' cams due to their appearance (Figure 12.4). Lost action cams are employed when a core pulling delay is required during the mould opening phase. In this case the parallel part of the pin enables the mould to move apart without initially moving the side core blocks. Movement of the side core components only starts when the mould has opened sufficiently enough to engage the angled section of the cam pin within the side core block hole. Once engaged, the cam pin actuates the block movement like any normal cam pin pulling sequence. Lost action cam pin assemblies are typically employed when moulded undercut features are required to retain the moulding on a desired half of the mould, usually for ejection purposes. The increased length of the dog-legged pins, compared to conventionally straight cam pins, means they are more easily damaged if abused in use. Additionally, deeper mould bolsters have to be employed to house the extra pin lengths resulting in a reduced working daylight once set in the moulding machine.

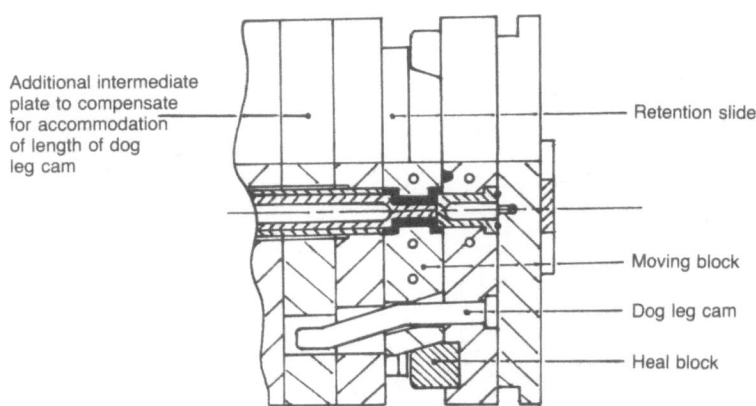

Additional intermediate plate to compensate for accommodation of length of dog leg cam

Retention slide

Moving block

Dog leg cam

Heal block

Figure 12.4 Lost action cam pin assembly.

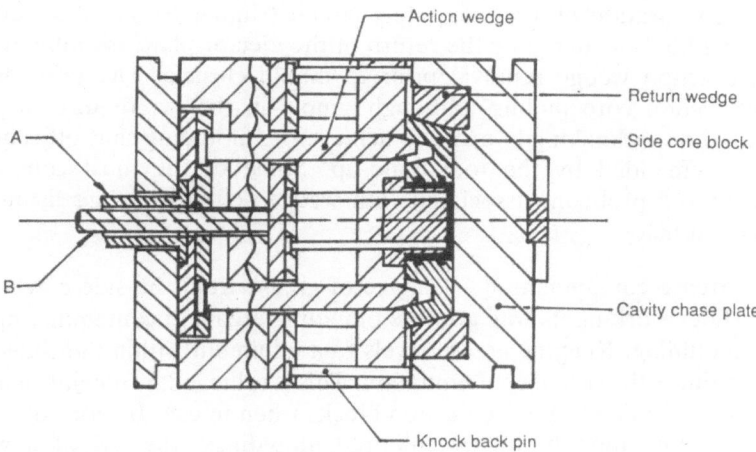

Figure 12.5 Typical wedge action assembly. A: First stage ejection (side core parting). B: Second stage ejection (component ejection).

12.3.3 *Action wedges*

The wedge actuation technique (Figure 12.5) is frequently used when there is a need for minimal side core movement with which to clear the moulding. Moulded features which include textured surface finishes, reverse draft tapers and shallow coring features may be produced using this pulling actuation method. The wedge assembly comprises four basic components:

(i)   the action wedges;
(ii) the side core block;
(iii) the block slides;
(iv) the return wedge.

The actuation force is applied to the tapered underside of the side core block by the action wedges, forcing the block away from the mould centre core. The action wedges, being retained in the mould ejector plate assembly, are essentially part of the ejection system. In order to enable the side core blocks to move prior to the ejection of the moulding, the action wedges tend to be longer than their accompanying ejector pins. After mould opening, wedge actuation and component ejection, the side core blocks are returned to shut-off against the mould centre core. This is achieved by the pushing action of the return wedge against the tapered back of the side core block as the mould closes. Retraction of the mould ejection system must happen prior to mould closure if damage is to be avoided to the underside of the side core blocks and the action wedges.

The incorporation of a mould safety circuit (similar to that described in section 12.3.2) actuated by the return of the ejector plate assembly would ensure action wedge removal prior to mould closure. The principle of wedge action core pulling, although simple in theory, requires a good standard of toolmaking in order to achieve the moulding shut-off requirements demanded by the tool build-up. Some of the most commonly encountered problems associated with wedge action tools are listed and explained below:

(a) Grease contamination – leakage of grease from the side core block slides working its way into the moulding area and contaminating the moulding. Keeping grease levels to a minimum within the slide area reduces the risk of contamination. This is achieved by regular removal and cleaning of the side action blocks when in use. In cases of minor contamination of lightly coloured mouldings, the use of a white silicone-based grease may prove beneficial.

(b) Wedge and slide wear – wearing of the return wedge and the block slide faces will eventually result in a poor core shut-off and flash occurring. The addition of hardened steel wear pads, containing grease grooves, into the wear faces enables tool wear to be compensated for (by packing behind the pad), with the ability to remachine the contact faces.

(c) Mould seizure – seizure generally occurs in wedge action tools between the opening action wedges and the side core blocks themselves. Lack of grease causes frictional overheating to occur. This dries out the remaining lubricant resulting in scuffing and seizure of the moving parts. The addition of grease grooves to the sides of the action pins will reduce this problem. However, in some cases the choice of steel type and hardness has a bearing on the wear resistance of the moving parts.

12.3.4 *Hydraulic core pulling*

Most modern moulding machines are hydraulically powered but not all machines are fitted with auxiliary hydraulic circuits. In cases where spare auxiliary circuits are available, the mould can utilise them for core pulling purposes. Figure 12.6 illustrates a simple hydraulically actuated pulling system, in which pressurised hydraulic oil is used to move the cylinder piston. Forward movement of the piston pushes the linked side core block inward against the centre core shut-off. Outward movement of the core block is achieved by diverting the pressurised oil flow from the top of the piston crown to the crown underside. This is usually achieved by means of a servo valve actuated by the machine control system. While the side cores are in contact with the centre core during the filling and cooling

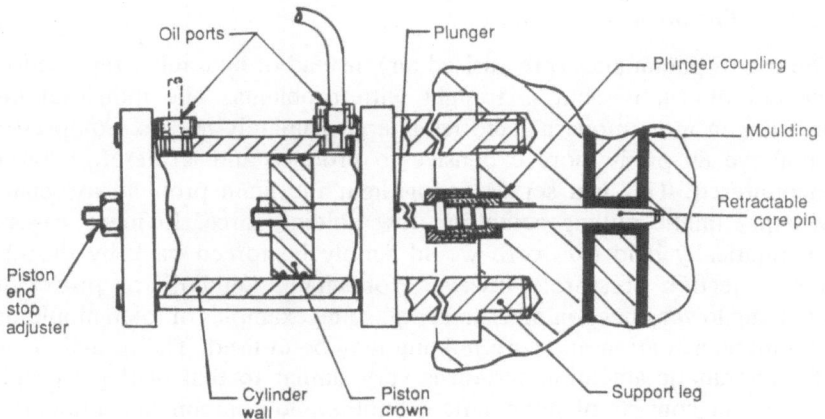

Figure 12.6 Hydraulic pulling actuation.

stages of the mould cycle, an oil pressure loading is maintained. Sufficient pressure should be applied to avoid the side cores being forced back by the increasing injection force as the impression fills. As a result of the auxiliary nature of this actuation method, mould safety must be adequately catered for if damage is to be avoided. Most damage is usually inflicted by the inward actuation of the pulling circuit with the mould open. Subsequent closure of the mould often results in core breakage or shut-off damage. Mould safety circuits (section 12.3.2) on hydraulically actuated moulds should not rely on a single micro-switch actuated in the core back position. In such cases, two switches should be employed: one at the side core block front position, the other in the conventional rear position. A system used in this manner is usually referred to as a tandem safety circuit. The signals generated are used to control the opening and closing of the moulding machine.

The principal advantages of using hydraulics as a core pulling actuation method are:

(a) Loading forces can be easily controlled and adjusted while in use;
(b) High closure and holding loads can be generated and maintained.

The principal disadvantages of hydraulic actuation include:

(a) Movement damping (deceleration of movement) can prove to be difficult to achieve at higher loading pressures;
(b) Hydraulic systems tend to be messy and are prone to contaminate their working locality.

12.3.5 *Pneumatic core pulling*

The use of pneumatics (pressurised air) instead of hydraulics represents a cheaper alternative but is fraught with problems. The industrial line pressure in most modern factories is approximately 80 psi. Compressed air above 80 psi is more expensive to produce and is therefore rarely encountered. This fact serves to highlight the main problem associated with pneumatic pulling actuation: low holding force. In many cases a pneumatically held side core would simply be forced back by the first stage injection pressure. However, for small mouldings or processing involving lower injection pressures, e.g. some examples of foam moulding, pneumatic actuation and core holding may be utilised. The design layout of a pneumatic actuation system is very similar to that of the hydraulic system but consists of pneumatic cylinders and components. Pneumatic components are available from a wide range of suppliers and can be replaced as standard mould parts as and when required. The same mould safety considerations are equally relevant with pneumatic actuation as they are with hydraulic actuation.

The principal advantages of employing pneumatics as a core pulling actuation method are:

(a) Speed of actuation and movement – pneumatically actuated components are fast and efficient, enabling fast cycling if required.
(b) Cleanliness – when supplied with filtered or 'scrubbed' air, pneumatics may be used for 'clean room' environment applications, e.g. medical moulding and robotics.
(c) Cheapness – pneumatic components are relatively cheap to obtain and maintain while in use.

The principal disadvantages of employing pneumatics as a core pulling actuation method are:

(a) Low holding force once applied.
(b) Noise – due primarily to the exhausting of actuation cylinders when in use. Pneumatic noise tends to be high pitched and is therefore frequently over the safe working threshold limit for unprotected hearing.
(c) Moisture – condensation may appear within the factory compressed air supply and is most pronounced if the compressor is sited outside the factory main building. The installation and use of condensation traps along supply lines help to reduce the moisture problem. Condensation may also appear within pneumatic cylinders buried within mould plates, especially in cases where chilled water is being used to cool the mould. Corrosion and piston sticking may result if pneumatic piston chambers and pipes are not cleared after mould use.

(d) Pressure fluctuation – pneumatic line pressure is invariably subject to fluctuation during the course of a typical factory working day as a result of changing demand. Pressure drops tend to slow actuation speeds and reduce holding pressures whereas pressure increases result in the opposite effect.

12.3.6 *Electro mechanical core pulling*

Electric motors are frequently employed to actuate and power core unscrewing mechanisms. In most cases power from the motor is transmitted via a gear drive system to the subject core(s) to be rotated (Figure 12.7). Component unscrewing is generally achieved with the mould in the closed position. This is to ensure that the moulding does not move or rotate while the thread is being disengaged. To enable rotation, the core body is supported on bearings either side of the keyed drive gear. The core body, being freely supported, is allowed to travel back into the mould as the core thread is unwound from the moulding. Once unwound, the electric motor is deactivated by a micro-switch positioned behind the moving core. The mould is then opened and the moulding ejected conventionally. After ejection the mould is closed and the cores moved forward for the next shot. The process is repeated.

Apart from the electro-mechanical unscrewing method previously mentioned, many other unscrewing techniques are commercially employed.

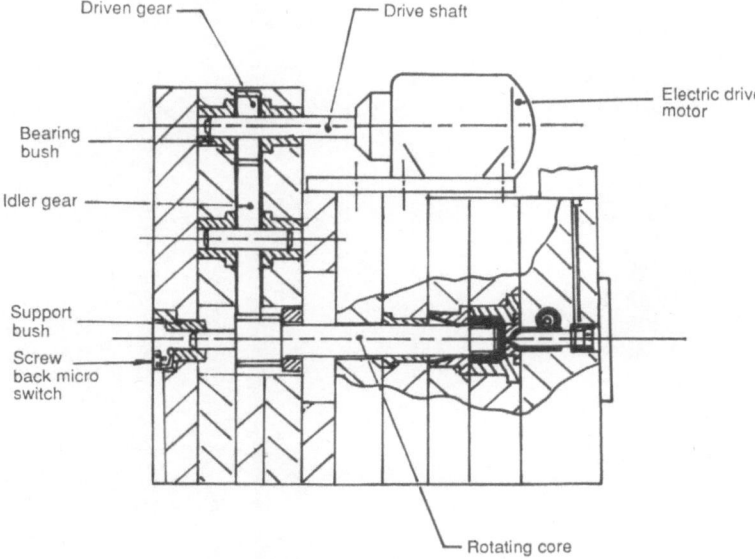

Figure 12.7 Electric unscrewing mechanism.

One of the most commonly employed non-electric unscrewing devices is based on a 'fast' helix screw thread actuated by the opening action of the mould. Such mechanisms can be obtained from catalogue-order mould part suppliers and fitted as standard mould components. Whichever unscrewing device or mechanism is employed, attention should be given to providing an adequate mould safety system in order to adequately protect the tool when in use.

13 Standard mould parts

13.1 Introduction

Standard mould parts are present in most modern mould tools in one
form or another. The days of manufacturing all the constituent mould
components in-house are long since gone: the practice almost stopped
due to the decline of the large in-house toolrooms. In most modern
factories the prime function of the toolroom tends to be mould maintenance
and production support; new moulds are generally manufactured off the
premises by more financially competitive 'trade' toolmakers. A whole
industry has been created which manufactures and supplies standard
mould parts based on a catalogue selection and order system. Fast order
delivery times (usually 24 h) are one of the major attractions to the
toolmaker, an important factor if the mould has broken down and is
required for production.

13.2 Why use standard mould parts?

Apart from the fast delivery service operated by the part suppliers, other
less immediately recognisable user advantages exist, which include:

(a) Shorter mould construction lead times. Figure 13.1a illustrates how
the mould construction time is typically apportioned (in % working
hours) for a mould constructed from non-standard mould components.
Figure 13.1b illustrates how the construction time is typically apportioned for the same mould, but this time incorporating standard
'ready-to-use' mould components.
Figure 13.1 visually illustrates the effect of using standard mould
components on the production of a typical mould tool, the most
significant effect being the increase in available production time for
the manufacture of the core and cavity components, resulting in
shorter manufacturing lead times.

(b) Increased machining capacity. The resultant increase in machine
tool capacity may be directed toward the manufacture of the 'rough'
tool work, e.g. machining insert pockets, side action blocks, etc.,
fully utilising the capacity available. In some companies this increased
capacity has been replaced with more specialised machines for the

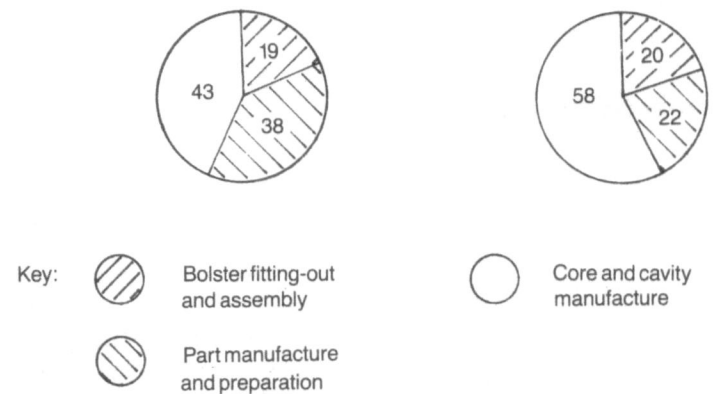

Key:

⧄ Bolster fitting-out
   and assembly

◯ Core and cavity
   manufacture

⧅ Part manufacture
   and preparation

Figure 13.1 Mould construction time apportioned in % hours for (a) a mould constructed from non-standard components, (b) the same mould constructed from standard 'ready-to-use' components.

manufacture of core and cavity components, e.g. specialised eroding machines, digitised CNC machining centres, etc. In other words, a modern mould-making company has, to some extent, become equipped for the use of standard mould parts.

(c) Ease of tool maintenance. The overall cost of mould maintenance should be split into two sections:

(i) The actual cost of maintaining the mould, i.e. replacing worn or damaged components.

(ii) The cost to the company in lost production from the mould.

The use of standard mould components has the effect of simplifying mould maintenance requirements and reducing tool downtime to a more acceptable level. This has the effect of reducing the overall long-term costs. The main disadvantages concerned with the use of standard mould parts tend to be associated with the following:

(a) Reliance upon one specific source of supply.

(b) Failure to supply on time, a rare occurrence with most part companies, but should be considered before adopting any product range.

(c) Mistakes in part ordering, by either party, are costly with respect to the increased mould downtime created.

13.3 Standard parts and assemblies

Standard mould parts are divided into two general groups:

1. Standard mould components or elements, i.e. the general fabric of the tool, e.g. the bolster itself, pillars and bushes, main plates, etc.

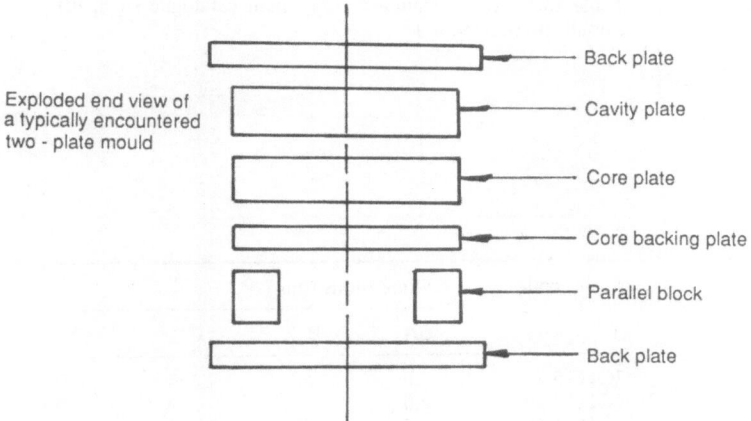

Exploded end view of
a typically encountered
two - plate mould

- Back plate
- Cavity plate
- Core plate
- Core backing plate
- Parallel block
- Back plate

Figure 13.2 Standard bolster elements.

2. Accessory components, usually concerned with mould cooling, feeding, core pulling or ejection. Accessory components include: ejector pins and sleeves, mould heaters and thermocouples, spiral cooling sleeves, standard core pins, etc.

13.4 Standard elements

Figure 13.2 illustrates the principal elements which make up a typical standard element mould bolster. Each element carries its own identification code number which identifies the range from which the part is a member and the material with which it is manufactured. The dimensions and steel grade for each specific element may be varied to suit the end use requirements and ordered accordingly. This system of assembling a mould from standard elements is known as the modular approach to mould construction.

Care should be taken when reading and ordering from standard part stock lists as mistakes are easily made. Table 13.1 is typical of stock lists encountered from some standard part manufacturers. A typical part order code is usually built up of groups of numbers and letters, each group denoting a different feature of the part being ordered. The individual order code groups usually cover the following part features:

(a) the material or steel type from which the part is manufactured;
(b) the size of the feature, if specifically required;
(c) the standard part group identity prefix.

Table 13.1 Typical standard component catalogue stock list (mould thermocouple)

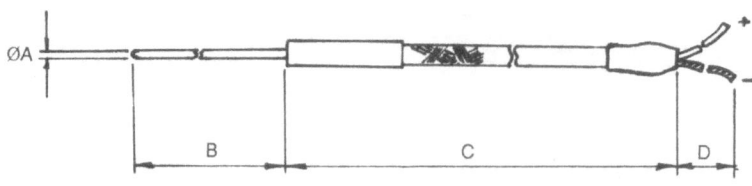

Order code	Dimensions (mm)			
	φA	B	C	D
ICTC-5-15	2.0	50	150	100
ICTC-5-30	2.0	50	300	100
ICTC-5-60	2.0	50	600	100
ICTC-10-15	2.0	100	150	100
ICTC-10-30	2.0	100	300	100
ICTC-10-60	2.0	100	600	100

ICTC: Iron constantan thermocouple

13.5   Accessory components

Standard part accessory components are generally incorporated into a mould design to reduce the volume of routine machining undertaken during mould manufacture. Broadly speaking, accessory components fall into four main mould component categories:

(a)   cooling components;
(b)   ejection components;
(c)   mould feeding systems;
(d)   core pulling parts and accessories.

Apart from the four main accessory groups listed many other tooling applications exist for which mould accessory components are available.

13.5.1   *Accessory cooling components*

Accessory components for cooling purposes usually take the form of bushed sleeves or jackets which either fit inside or about the feature to be cooled. A spiral flow path is machined into the jacket through which coolant is circulated. Rubber sealing rings are used to create a watertight seal between the cooling sleeve and the mould components (chapter 8) which require periodic replacement during mould servicing. For mould

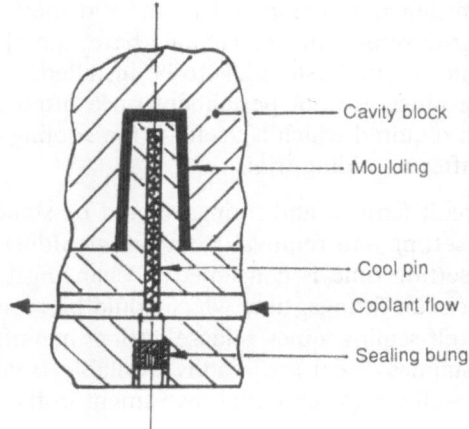

Figure 13.3 Cool pin cooling.

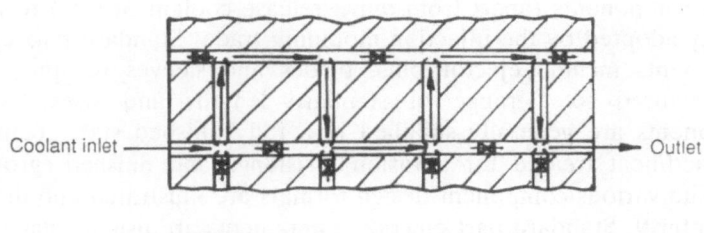

Figure 13.4 Use of sealing plugs.

features which are too small for the inclusion of a 'bubbler' (section 8.3.6) or the jacketed/sleeved cooling techniques discussed previously, accessory 'cool' pins or rods are frequently employed.

The principle of the cool pin cooling technique (Figure 13.3) relies on the thermal conductivity of the pin itself and that of the coolant circulating about the pin end (chapter 1). For maximum thermal efficiency cool pins should be kept as short as practically possible and should be tightly fitted into their respective drilled holes inside the mould feature. Cool pins are not as efficient as a locally circulating coolant, but are usually the only remaining option available for very thin core and cavity features.

In recent years many specialised accessory components have become available from the standard part manufacturers which include self-sealing plugs or bungs (designed on the expanding collet principle) used to create cooling circuits within the mould (Figure 13.4). The principal reasons for using expanding plugs are:

(a) Reduced fitting time compared to the old method of tapping and sealing a grub screw into the channel bore, a major consideration if a large-scale cooling system has to be installed.

(b) Expanding plugs are not permanent once fitted and may be repositioned as required which is useful if the cooling circuitry has to be modified after moulding trials.

The cooling circuit fixtures and fittings should be standardised to enable ease of mould setting and removal from the moulding machine. A large percentage of setting time is consumed in changing and connecting the mould cooling circuit fittings, time which could be more profitably spent. Standard part, self-sealing 'quick release' fittings manufactured from brass or better still stainless steel are readily available from many sources of supply and are well worth the initial investment in the long term.

13.5.2 *Accessory ejection components*

Ejector pins and associated components were probably the first standard mould components (apart from quick release coolant fittings) to be universally adopted by the injection moulding trade. Standard part ejection components include ejector pins, blades and sleeves (chapter 9), all manufactured to a range of standard lengths and sizes. Ejection components are generally supplied in a fully finished state, being case hardened/heat treated for abrasion resistance and finished, ground to size. The various component design formats are illustrated and discussed in chapter 9. Standard part ejection components are usually classified by their type, section profile and critical dimensions. Figure 13.5 illustrates a typical ejector pin. The part classification reads: pin round, DIA. 4 mm, L 150 mm.

The good dimensional accuracy and the relative cheapness of these components has led to their widespread adoption throughout the mould-making industry as standard mould components.

13.5.3 *Mould feeding systems*

Mould feed systems, principally hot-runner feed systems, are almost totally constructed from standard components. The hot runner market is

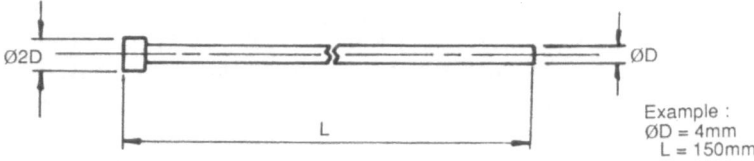

Figure 13.5 Ejector pin.

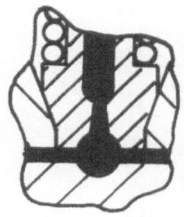  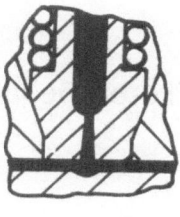

Figure 13.6 Use applications of a heated sprue brush.

virtually saturated with different designs and types of feed systems. Choosing the correct system for the intended moulding application is a difficult business (chapter 11). When choosing a hot runner feed system additional consideration should be given to part serviceability and availability as standard components. Replaceable components which include cartridge heaters, band heaters, thermocouples and heater controllers are often available from alternative suppliers at reduced cost for mould servicing requirements. Apart from complete hot runner feed systems, many other individual components are available which include conventional sprue bushes, heated sprue bushes, hot tipped gates and nozzles. The comprehensive range of mould feeding accessory components offered by the standard part manufacturers enables the mould designer to specify accurately the intended use requirement. Examples of use applications based on one accessory component are illustrated in Figure 13.6. The subject component in this case is a heated sprue bush.

13.5.4 *Core pulling parts and accessories*

Standard part manufacturers have spent considerable time and effort developing core pulling mechanisms, some for specific applications, e.g. thread disengaging. The main advantages to the mould designer when specifying a standard part pulling device are:

(a) The standard part designs are proven and known to work.
(b) The fitting and working dimensions are freely available without the need to calculate the working movements.
(c) Unit costs are known from the onset making costing and tool estimating more accurate to forecast.

Core pulling methods are illustrated and discussed in detail in chapter 12. A rule worth noting when selecting a core pulling technique must be 'simplicity is the best policy'. This rule is also applicable to the mould safety method used in conjunction with the chosen pulling technique.

14 Prototype moulds

14.1 Introduction

Prototype moulds are constructed and used specifically to obtain information about the component to be moulded. Mould construction varies from multi-impression pilot production tooling, based upon the final tooling quality and construction layout, to single impression aluminium or epoxy two-plate mould inserts at the cheaper end of the scale. Choice of prototyping tool design and construction method usually depends on the significance of the component to be moulded and the amount of processing information required.

14.2 The case for a prototype mould

The employment of a prototype mould yields considerable information about the component to be manufactured and about the design requirements of the final production mould. Therefore, it is possible to divide the case for a prototype mould into two specific areas. One area is concerned with the moulded component and the other with the mould design and its constructional aspects.

14.2.1 *Moulding aspects of the prototype component*

The moulding performance information yielded about the prototype component will include the following:

(a) *Dimensional stability of the moulding after manufacture*. Dimensional instability within an injection moulded component is usually caused by a combination of contributing factors which include:
   (i) Differential shrinkage – usually caused by thick to thin sections of polymer within the component or due to the aggravated effects of internal orientation of the polymer about certain component features.
   (ii) The inclusion of stress raising features within the component design itself. Stress raising features include sharp 90° corners, raised bosses and webbed thick to thin section interchanges. In fact, stress concentrations may appear wherever the flow of

the polymer is interrupted or restricted within the mould impression.

(iii) The effect of certain moulding conditions, primarily mould surface temperature and cooling time, have a marked effect on the component after-moulding dimensional stability and shrinkage. As a general rule, 'hot' moulds coupled with short cooling times increase post moulding shrinkage and dimensional instability.

(b) *Component aesthetic considerations.* Certain aesthetic considerations become apparent only when a moulding is produced, hence the importance of a prototype mould. The visual appearance of the moulding is often of special importance. Examples would include cosmetic packaging and automotive decorative trim. The aesthetic factors most likely to become apparent by prototype moulding include:

(i) the position and prominence of weld or meld lines;
(ii) the location and visual appearance of gate scars;
(iii) the effects of poor cavity venting, i.e. surface discoloration or burning;
(iv) the prominence and the likely location of flash;
(v) effect of the component surface finish.

The surface finish of the component can be assessed with respect to its moulding practicality, e.g. rough or course spark eroded finishes may be tested to see if additional side drafting of the component is necessary. In many cases, especially when moulding reinforced grades of polymer, a coarse finish might be preferred to a smooth finish.

(c) *Quality control aspects.* The process of injection moulding can be considered as a variable sequence of events, each and every moulding cycle. Production variances will exist on a minor scale which will directly affect the quality of the moulded component, both dimensionally and aesthetically. The size and magnitude of the production variance may be assessed by undertaking prototype moulding trials. The results can be used to construct a production QA specification. Failure to adequately assess the production QA aspects of a job could prove to be a costly long-term mistake in terms of tooling modifications and production rejects.

14.2.2 *Mould design and constructional aspects highlighted by prototype moulding*

The working limitations of the prototype mould design are usually exposed by the overall quality of the moulding produced. The principal mould design faults exposed by prototype moulding are listed and discussed below:

(a) *Localised hot or cold spots within the core and cavity build-ups.*
These are often associated with the design and efficiency of the
mould cooling system. The effects of core or cavity hot/cold spots
usually result in uneven filling of the impression leading to gas
entrapments, component burning, uneven shrinkage or localised
distortion.

(b) *Mould overheating.* Caused by an inadequate mould cooling system
or the use of the incorrect cooling medium. The design of the
moulded component may not allow for the incorporation of an
adequate mould cooling system. The effects of excessive mould heat
gain generally result in the following processing and quality related
problems:
    (i)   component dimensional instability;
    (ii)   component sticking in cavity;
    (iii) component ejection damage;
    (iv)   component warpage and distortion;
    (v)   increased cycle times;
    (vi)   inconsistent production quality.

(c) *Ejection inadequacy.* A poorly designed and located component
ejection system will always result in moulding problems. For example,
the use of ejector pins (chapter 9) against thin or unsupported
component wall sections may lead to component distortion, damage
or increased levels of internal component stressing, leading to long-
term post mould reversion. Increased cycle cooling times and
component damage are the most probable results of an inadequately
designed and located mould ejection system.

(d) *Efficiency of the mould feed system.* The design and efficiency of the
prototype mould feed system are of great importance, with respect
to:
    (i)   the quality of the moulded component;
    (ii)   the final design and construction layout of the intended pro-
           duction mould tool.
Inadequacy of a mould feed system may appear in a number of
ways, which include:
    (i)   High injection pressure requirements, usually due to undersize
           feed runners or the incorrect design of adopted gate (chapter
           7). A prototype mould provides an ideal situation with which
           to size and match a feed system best suited to the intended
           purpose.
    (ii)   Protracted cooling times, especially if thin-walled mouldings
           are being produced. In such cases the size and the cooling
           (freeze) time of the feeding runners and gates could prove to
           be the major contributing factors toward the total cycle cooling
           time requirement.

(iii)   The effects of unbalanced feeding and mould filling. The gate
        location and geometry dictate the manner in which the incoming
        molten polymer fills the impression. This determines the filling
        balance of the moulding. The resultant mouldings produced
        could suffer from any combination of the following quality
        faults:
        − localised under or overpacking, resulting in uneven shrink-
           age and distortion;
        − poor or uneven component surface finish;
        − gas traps, voids and surface burning of the moulding;
        − weak weld lines;
        − Intermittent component flashing and shorting;
        − inconsistent component shot weight.

14.3   Prototype mould tool materials and construction

The choice of prototype mould construction method and materials is
dictated by the design and trial requirements of the subject component.
As in the cases of the two-plate and three-plate mould construction, the
materials chosen for the prototype mould should reflect the end use
application. Table 14.1 suggests the use of specific mould construction
materials for the listed prototype moulding applications.

Table 14.1 Use of specific mould construction materials

Prototype moulded component	Mould construction material	Reason for selection
28 mm dia. HDPE bottle closure	Mould tool steel e.g. AISI P20. Case hardened core and cavity components	Mould must perform trial production runs
Rigid PU foam computer case	Epoxy resin	Low moulding forces, component design and aesthetic details of prime importance
PS models/toys	Aluminium alloy bolster, core and cavity components	Ease of machining modification. Large cheap bolsters
Small components	Steel inserted bolster, BeCu core and cavity inserts	Standard cooling and ejection in bolster. BeCu good thermal conductance. Ease of modification

14.4   Prototype tool construction

The choice of prototype tool construction method is usually dictated by the complexity of the component design and the tooling budget available. In general, the most commonly encountered approaches usually include:

(a)   conversion of a suitable obsolete mould tool bolster;
(b)   utilisation of the inserted bolster design format (chapter 6);
(c)   adoption of a modular tool construction system.

14.4.1   *Inserted bolster prototyping*

The use of an inserted bolster (chapter 6) mould tool design for prototyping purposes offers the user some distinct advantages over other tool construction methods. In terms of prototyping use, insert interchangeability, reduced tooling and construction lead times are the main benefits to the end user. The prototyping limitations usually encountered with this design of mould are:

(a)   poor cavity and core cooling;
(b)   inadequate positional location of the ejection system;
(c)   reduced size requirement of the moulding to be prototyped.

In most cases, a standard size two-plate steel bolster is employed for prototyping purposes, usually fitted with oversized core and cavity plates (Figure 14.1). The depths and dimensions of the machined insert core and cavity plate pockets dictates the size and number of impressions able to fit the bolster. Cooling channels are positioned and drilled about each insert pocket. Sufficient space should be allowed between the channel locations to enable the mould ejection and feeding systems to be incorporated. An

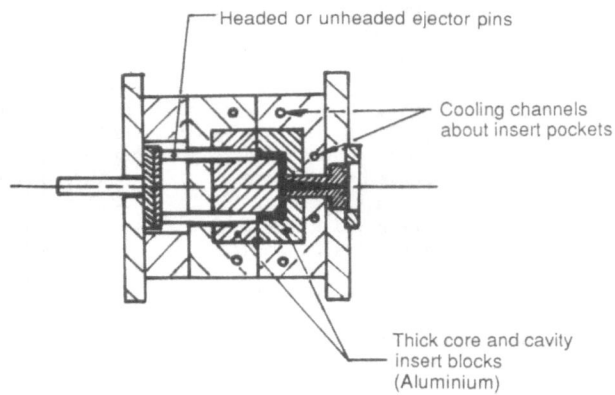

Figure 14.1 Inserted bolster prototyping.

additional spacing allowance should be included for modification purposes if required. The inclusion of an extended sprue bush or a heated nozzle enables feeding of the prototype impression. Rather than machining the sprue feed system directly into the solid insert block, the use of a bushed sprue feature eases insert modification, cleaning and changing when in use.

Core and cavity detail can either be inserted into the insert blocks or machined directly to form depending upon the component complexity or its use requirements. Due to the relative isolation of the core and cavity blocks from the main mould cooling system, many toolmakers prefer to use materials of superior thermal conductivity for the mould core and insert components, two of the most commonly employed materials being beryllium copper (BeCu) and aluminium based alloys. This concept has been taken a step further by certain companies in recent years which choose to manufacture their production mould tools in such a manner. In many cases cooler and more efficient coolants are required, e.g. liquid carbon dioxide or pressurised sub-zero glycols, for increased cycling control and mould cooling efficiency. For prototyping purposes, inserted bolster design tools generally rely upon pin, sleeve or blade ejection systems, providing ease of actuation and incorporation into the insert design.

In recent years the appearance and use of vacuum hardenable mould tool steels has increased the adoption of the inserted bolster design for production and prototyping purposes. Initially the soft (unhardened) steel is used to manufacture the prototype core and cavity insert components. Once trialled, modified and proven to an acceptable standard, the soft

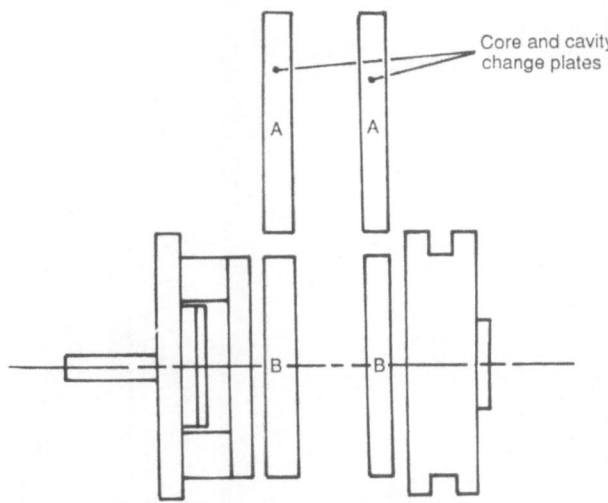

Figure 14.2 Modular mould tool construction.

inserts are vacuum hardened to the required working hardness to inhibit oxide/scale formation on the surface. This approach has proved to be popular if an element of design risk is attached to the initial prototype component design.

14.4.2 Modular prototype mould tool construction

The modular or 'change plate' prototyping approach is usually employed in cases where the established production tooling design permits plate interchangeability. The plates concerned tend to be the core and cavity plate assemblies and in some cases an additional stripper plate. In general, purchased standard mould plates are utilised (standard part numbers should be included on all detail drawings if change parts are to be used). The prototype core and cavity features are either directly machined into the plates themselves or an inserted format may be employed (section 14.4.1). The parent mould's feeding, cooling and ejection systems are usually deliberately utilised by the prototype plate assemblies during the mould tool construction. Figure 14.2 shows a typical modular construction.

This approach to prototyping is generally employed for multiple impression tooling trials, i.e. when similar components are to be prototyped within an existing product range. Although more costly than other prototyping techniques, the best results are usually achieved in terms of component quality, consistency and moulding performance. Typical applications vary from large individual mouldings, e.g. foamed casings and bodies, to the smaller, more typical, volume produced mouldings, e.g. ranges of bottle caps and closures.

15 Mould tool materials

15.1 Introduction

The choice of mould construction material is of great importance if the mould is to function effectively. The prime function of any mould construction material must be that of meeting the service requirements imposed upon it. Moulding requirements vary, e.g. from simple prototype work undertaken on soft aluminium construction moulds, to fully hardened alloy steel volume production moulds moulding to close dimensional tolerances. In order to avoid costly long-term mistakes, the tooling application must be thoroughly investigated and fully understood before a selection decision can be made. Once decided upon, the selected material should be written into the tooling specification for future procurement purposes. The use of a tooling specification laying out the mould requirements prior to tool quotation or construction greatly reduces the risk of fundamental mistakes being made at the toolmaking stage.

15.2 Mould construction material requirements

A mould construction material should possess qualities or attributes relevant to the intended application. For general purpose moulding applications, the principal material attributes are listed and explained below.

(a) *High core strength*. As a result of the service conditions encountered during injection moulding, i.e. relatively high compressive cyclic loadings, the material core strength is of relevance to the mould designer. The material must be able to withstand high compressive loads without cracking or splitting.

(b) *Good wear resistance*. Mould tools are subject to considerable wear from many sources, which include:
  (i)   the polymer itself;
  (ii) the mould ejection system;
  (iii) the wearing action of shut-off faces;
  (iv) abuse during cycling and shut shots.
  Wear resistance may be imparted to a mould tool material by various means, usually by hardening the material or the addition of property modifying alloying elements. The choice of which method depends on the material in question.

(c) *Excellent surface finish*. A good serviceable surface finish is of the utmost importance, especially for core and cavity components. The intended material should be capable of sustaining a good long-term surface finish without the additional requirement of polishing between production runs. As with wear resistance, the material's surface hardness and composition have the greatest influence on its finishing properties.

(d) *Dimensional stability*. The cyclic loading nature of the injection moulding process subjects the mould materials to considerable levels of stress and elastic deformation. The ideal mould material should possess sufficient strength and durability to resist permanent deformation but sufficient ductility to resist cracking and impact loadings. For this purpose, many grades of mould materials, especially steel alloys, have been developed to fulfil the above requirements.

15.3 Mould construction materials

Broadly speaking, mould construction materials can be divided into two groups: ferrous and non-ferrous materials.

15.3.1 *Ferrous mould construction materials*

For general purpose injection moulding applications, alloy steels are usually employed instead of standard plain carbon steels. The main limitations associated with plain carbon steels, with respect to an injection moulding application, are:

(a) High strength and abrasion resistance are achieved by increasing carbon content, which when heated and rapidly cooled results in steel embrittlement leading to distortion and cracking.

(b) Poor corrosion resistance. A plain carbon steel would rust easily if used in a typical injection moulding environment.

(c) Inability to maintain a polished surface finish. A plain carbon steel would tarnish readily under normal moulding conditions if employed for the manufacture of core or cavity components.

15.3.2 *Alloy steels*

The term alloy steel describes a steel material which contains other alloying elements in addition to carbon which have been added to deliberately modify the properties of the steel. Mould tool steels contain quantities of specific elements which combine with the steel to improve the material's intended application properties.

Table 15.1 Typical alloying elements

Element	Typical amount (%)	Main effect on properties
Aluminium Al	0.5–1.3	Aids nitriding[a]
Chromium Cr	0.5–2.0	Increases hardenability and improves corrosion resistance
	4.0–18.0	
Manganese Ma	0.2–0.4	Acts with sulphur to reduce brittleness
Molybdenum Mo	0.1–0.5	Aids heat treatment and improves strength and toughness
Nickel Ni	0.3–6.5	Increases hardenability and improves strength and toughness
Silicon Si	0.2–2.0	Increases hardenability and limits oxygen in steel making
Tungsten W	1.0–18.0	Improves hardness at higher working temperatures

Other alloying elements which are often added to an alloy mould tool steel include sulphur or lead (improves machinability) and vanadium (improves strength and toughness).
[a] Nitriding – surface heat treatment process used to increase the surface hardness of steels.

Typical alloying elements used in mould tool steels and their contributing effects upon the alloy produced are listed in Table 15.1.

The alloying elements and their addition percentages vary according to the different steel manufacturers and the intended applications. Choosing a steel type for a mould tool application is a difficult business, especially for the beginner. Most steel producers manufacture and supply mould tool steels bearing their own trade names and designation code numbers. Within the steel industry, various organisations and institutions exist which are responsible for the quality and national standard of steels. Some of the more prominent are:

(a) American Iron and Steel Institute (AISI);
(b) Swedish Standardisering Kommission (ISI);
(c) various European standards institutes, e.g. DIN and BS 970.

Internationally the AISI designation has been adopted and most of the major steel manufacturers give equivalent designation code numbers. An AISI code reads, for example, AISI D3, where D denotes the class of steel, and 3 denotes the grade of steel within the class. In this case AISI D3 is a high carbon content, high chromium steel used for mould core or cavity components.

15.3.3 *General purpose mould steels*

As a result of the demanding requirements of mould steels, alloy steels are commonly used during the construction of mould tools and dies (see section 15.3.2).

The AISI class designation for mould steels is the letter 'P' for general

Table 15.2 Composition and heat treatment details for AISI P class steels (Source: *Machinery's Handbook*)

	Mould steel type (AISI)						
	P2	P3	P4	P5	P6	P20	P21
Elements (%)							
C	0.07	0.10	0.07	0.10	0.10	0.35	0.20
Al							1.20
Mo	0.20		0.75			0.40	
Cr	2.00	0.60	5.00	2.25	1.50	1.25	
Ni	0.50	1.25			3.50		4.00
Hardening temp. °F	1525 to 1550[a]	1475 to 1525[a]	1775 to 1825[a]	1550 to 1600[a]	1450 to 1500[a]	1500 to 1600[a]	soft
Tempering range °F	350–500	350–500	350–500	350–500	350–450	900–1100	aged
Tempered hardness R_c	64–58[b]	64–58[b]	64–58[b]	64–58[b]	61–58[b]	37–28[b]	40–30

[a] After carburising.
[b] Carburised case.
Carburising is a surface hardening process which involves changing the composition of the steel's surface layers. The process increases the steel's surface carbon content from less than 0.2% to about 0.7–0.8%. The resultant steel after quenching treatment has a harder surface layer (e.g. 60 Rc) and a softer, more ductile inner core (e.g. 20 Rc). The depth of the carburised case layer is dependent on time, temperature and the carburising medium employed.
Other surface hardening treatments exist which could also be employed, such as nitriding and carbon nitriding. For steels with a carbon content of over 0.4–0.7%, selective surface heat treatments are usually employed, e.g. flame hardening or induction hardening.

purpose mould tool applications. Table 15.2 shows composition and heat treatment details for AISI P class steels.

The AISI P range of steel grades possesses relatively good machining properties due to the alloy compositions. Machining processes such as spark erosion and all the usual metal cutting processes can be employed during tool manufacture. As a result of the Ni–Cr content of the steel, a good surface finish is achievable using various polishing and finishing techniques. The steel's corrosion resistance is also improved compared with a plain carbon steel.

While the AISI P range of steels represents a reasonable general purpose steel choice for moulding applications, extremes of application requirement will always exist. In such cases a more specialised steel should be selected, the properties of which reflect the use application intended. The following examples have been chosen to demonstrate the service conditions in which a specialised steel type/grade may be selected.

Example 1. When moulding or processing PVC, account should be taken of corrosive degradation of the material (HCl is a degradation product). The selected steel should possess good anti-corrosive properties; such steels usually contain higher Mo–Cr contents than usual. A chromium content as high as 18% may be required.

Example 2. Highly abrasive polymers which often contain mineral or fibre reinforcements should be adequately catered for when selecting tool steels for moulds and dies. Harder, more abrasion resistant steels, must be considered for mould cavity and core inserts. Alloy elements may include tungsten, cobalt, chromium or vanadium. Generally speaking, these harder abrasion resistant steels tend to be brittle in nature and are usually only used as inserts at high wear points within the mould, e.g. gates, core and cavity components.

Specialised steels, as described above, tend to be more costly to purchase and difficult to machine. For example high chromium or stainless steels are non-magnetic and can in fact burn if spark eroded. Tungsten carbide steels are extremely difficult to machine, blunting cutting tools easily and requiring much polishing and finishing to obtain a good quality moulding surface. In order to keep tooling costs down, high performance steels should only be specified when absolutely necessary.

15.4 Non-ferrous mould construction materials

Although the majority of mould tools are manufactured from alloy steels, a number of situations exist where the use of a non-ferrous material is more appropriate. Some application examples for selecting a non-ferrous material over a steel for a mould tooling application are suggested and discussed below.

(a) *Ductile use applications.* Thin mould core and cavity sections (especially where cooling channels are located) are subjected to high cyclic loadings, both from the operation of the mould and from the pressurised polymer itself. A toughened or even a soft steel could crack and collapse under such conditions because of its relative lack of ductility and the tendency of the material to stress crack under long-term cyclic loads, especially about thick to thin changes in section or from sharp corners. Variations in localised heat treatment can create areas of high stress concentration within a steel component resulting in a loss of uniform component ductility and an increase in the risk of component failure occurring. The use of beryllium–copper (BeCu) for thin sectioned core and cavity inserts is not uncommon in such situations. Other materials, which include aluminium or brass, are sometimes similarly used although aluminium/aluminium alloys tend to be too soft, whereas brass, like steel, exhibits a tendency to work-harden, if stressed in a cyclic manner.

(b)   *To increase thermal conductivity.* The thermal conductivity of some alloy steels is relatively poor in comparison to that of the beryllium—copper and aluminium based alloys (Table 8.2). In applications where adequate cooling is difficult to incorporate into the mould design (e.g. long thin core sections, raised webs, etc.) feature inserts manufactured from BeCu or Al are frequently incorporated into the mould build-up to facilitate faster heat removal from the immediate locality. Removal of excess heat from the inserted feature is usually achieved by extending the insert into the vicinity of a cooling channel deeper into the mould. This technique of heat removal has been adopted by some standard mould part manufacturers who refer to the conductive rods used as *cool* or *cold* rods.

(c)   *To provide 'bedding-in'.* The delicate nature of some mould features, e.g. thin shut-off cores or hard/brittle core shut-off faces, require excessive load protection. The use of a softer material for a shut-off face serves to protect the harder, more brittle, steel coring feature shutting down against it. This technique is also used to reduce the likelihood of flash propagation across a shut-off face as the tool beds-in with age, especially in cases where soft or only toughened core and cavity components are being used. The softer core shut-off face is replaced or re-dressed (machined flat) between moulding/production runs. The materials employed for this purpose are usually 'soft' aluminium or a 'softer' grade beryllium copper material.

(d)   For prototyping purposes. Materials other than steels are frequently employed, especially if an inserted design of bolster (chapter 6) is to be utilised. Materials such as epoxy and unsaturated polyester resins can be included in this category, usually in cases where 'concept' or foamed mouldings are required for initial assessment. Polymer-based resins are generally poor mould making materials because of their poor thermal conductivity. Moulds would overheat rapidly in use. Some companies nevertheless specialise in this form of prototype work using 'metal spraying' building up techniques to protect and finish the polymer core and cavity components for moulding purposes.

Bibliography

H. Gastrow (1983) *Injection Moulds*, Hanser, Munich.

R.G.W. Pye (1983) *Injection Mould Design for Thermoplastics*, 3rd edn., Longman, London.

A. Whelan (1982) *Injection Moulding Materials*, Elsevier Applied Science, London.

A. Whelan (1984) *Injection Moulding Machines*, Elsevier Applied Science, London.

J. Brydson (1990) *Handbook for Plastics Processors*, Heinemann Newnes, Oxford.

H. Morton Jones (1989) *Polymer Processing*, Chapman & Hall, London.

E. Oberg, F.N. Jones and H.L. Horton (1989) *Machinery's Handbook*, Industrial Press Inc., New York.

Index

ABS 2, 6, 13, 16, 29, 45, 84
AC heaters 91
accessory components 109
acetal 2, 6, 13, 16, 28, 29, 45, 84
acrylic 45
action wedge 101–102
aesthetic factors 57, 60, 61, 65, 77, 115
air insulation 90, 91, 92
alignment 54
alloys, non-ferrous
  aluminium 71, 117, 119, 123, 124, 125, 126
  beryllium/copper 60, 71, 92, 117, 119, 125, 126
  brass 71, 125
American Iron and Steel Institute 123
amorphous polymer 12, 13
anisotropy in mouldings 2, 4
antifreeze 71
apparent viscosity 28
arrestment notch 98
azo-carbamide 4

back plate 43, 47, 53, 60, 95, 109
ball catch 99
band heaters 113
bedding in 126
blind spots 92
block retention slide 98, 100
bolster 108
bosses 8–9, 114
British Standards (BS) 123
bubbler 111
bull nosed plunger 98
burn mark/burning 9, 116
bushes 108

CAD 9, 72
cam pin 97–100
  damage 99, 100
  misalignment 99
carbon dioxide, liquid 119
carbon nitriding 124
carburising 124
cartridge heaters 91, 92, 93, 94, 113
cavitation in mouldings 3
cavity
  mould 43, 53, 69
  plate 43, 53, 60, 87, 88, 91, 95, 109

cellulose acetate 2, 6, 16
chase plate 91, 101
chiller unit 21
clean room moulding 104
cold spots in mould 116
cold trap 57
cold well 62
colour change 86, 90, 94, 96
consistency factor 27
cool pins 69, 111
cool rods 111, 126
coolant flow rates 19–20
cooling 15–16, 54
  channel dimensioning 21–22
  channel location 72, 94, 96
  channel section 25, 26, 30, 31
  media 70, 71
  rate 16
  time 18–19, 69, 70
cooling systems
  accessory components 110–112
  baffle 75
  bubbler 74, 75
  cavity insert 74
  finger 74, 75
  parallel channel 73
  plane 73
  plateau 73
  series channel 73
  spiral channel 73
  spiral core insert 74
core 42, 43, 50, 69
  backing plate 43, 53, 81, 109
  insert 81
  plate 43, 53, 81, 87, 88, 91, 95, 109
  pulling 78, 97, 100, 113, 114
  strength 121
  unscrewing 105
corners in mouldings 8, 114
crystallinity in polymers 12, 13, 70
cycle times 18, 90

datum line 47
daylight clearance 78
DC heaters 91
dead spots 92, 94
degating 59, 61, 63, 65
densities of polymers 16
dimensional stability

dimensional stability *cont'd*
  mould   122
  product   114
dimensionless temperature   17
DIN   123
distributor
  block   93
  bore   93
  tube   93
dog leg   100
draft taper/angle   7, 77, 101
dragging   77

economic factors   55, 65, 66, 71, 80, 86,
  90, 93, 94, 96
ejection   46, 52, 69, 116
  accessory components   112
  actuation   78
  force   78, 84–85
  plates   43, 53
ejector, types of
  air   83
  bar   46, 47
  blade   46, 79–80, 112, 119
  pin   43, 46, 55, 64, 67, 79–80, 109, 112,
    119
  pneumatic   83
  sleeve   69, 80–81, 112, 119
  stripper plate   83–83
  stripper ring   82–83
  valve head   80–82
ejector plate   46, 47, 101
enthalpy   14, 15, 19
epoxy resin   117, 126
expanding plugs   112

feed plate   87, 96
ferrous materials   122–125
fillers   70
flash   37, 58, 78, 115, 126
flow equation   25
flow of polymer melts   27–28
flow path ratio   6, 7
fluorinated ethylene propylene   2
foam moulding   4, 104, 126
Fourier number   17, 18
freeze flow   56–57
freezing off   96
friction heating   102

gas entrapment   59, 64, 116
gas injection   4
gates   55, 58–66
  blockage   94, 95, 96
  bush   60
  diaphragm   61–62
  direct feed   59
  disc   61–62

  edge   59–61
  fan   61
  film   65
  flash   65
  pin   65
  pips   94
  position   58, 59, 69
  ring   62–63
  scar   64, 115
  spoke   83
  submarine   63–64
  tab   65
  tunnel   63–64
glass fibre   2, 16, 70, 96, 125
glycols   119
grease contamination   102

hardening   123
  temperatures   124
heal block   98, 100
heat
  soaking   92, 93
  requirements of polymers   13
  transfer   16–19
  treatment   54
heated probe   93, 94, 95, 96
heated sprue bush   113
heater unit   21
helix screw thread actuation   106
hot manifold block (HMB)   91–92
hot runner   87, 90–96, 112, 113
hot-runner block (HRB)   93–94
hot spots in mould   116
hot tip gate   113
hot tip nozzle   113
hydraulic clamp   37, 38
hydraulic core pulling   102–103

injection cycle   35–37
injection moulding machine
  base unit   34–35
  clamp unit   37–39
  configurations   39, 43
  injection unit   35–37
  nozzle heater   60
  operating cycle   36–37
insert bolster   52, 53, 118–120
insulation plate   95, 96
integral assembly   52, 53
interfacial vacuum   79
intermediate plate   87
internal stress   *see* moulded stress

jetting   65

laminar flow   24
latent heat   13, 16
limit bolt   88
liquid crystal polymers (LCP)   6

locating pillar   43, 47, 48, 53
locator bush   43
long fibre reinforced plastics   6
lost action cam pin   100–101

manifolds
  externally heated   91–92
  insulated   91, 95–96
  internally heated   91, 93–95
melt compressibility   32–33, 96
melt flow in mould   31
metal spraying   126
micro switch safety circuit   99–100, 102, 103
mineral reinforcement   2, 5, 16, 125
modular mould construction   120
molecular orientation   4–5
mould
  bolster   97, 100
  dimensioning   3
  feed efficiency   116–117
  feed system   112–113
  filling   31–33
  heaters   109
  impression   42
  material requirements   121
  overheating   116
  packing   36, 58
  seizure   102
  shrinkage   1–3, 114, 116
  shrinkage data   2
  size   54
  temperature   3
mould types
  family   49
  prototype   114–120
  runnerless   49, 90–96
  stack   49
  three-plate   48, 49, 59, 86–89
  two-plate   52–55, 59
  undercut   49
moulded stress   4–5, 36, 67, 114
moulding capacity   107
moulding cycle   107, 108
moving insert block   98, 99, 100

Newtonian flow   25, 26
nitriding   123, 124
nitrogen   4
non-ferrous mould materials   125–126
nylon 6   2, 6, 13, 16, 29
nylon 6,6   2, 6, 13, 16, 45, 84
nylon 11   2, 16
nylon 12   2, 16

oil
  heating   71
  contamination   103

coolant   21, 24, 25
opening plate   87
orientation   70

parallel block   43, 46, 53, 109
parting line   see split line
pillar bush   43
plate bush   83
pneumatic core pulling   104–105
Poiseuille's equation   21, 22, 26
Poisson's ratio   84, 85
polyarylates   6
polybuyleneterephthalate   2, 6, 29
polycarbonate (PC)   2, 6, 13, 16, 28, 29, 45, 84
polyester resin   126
polyetheretherketone (PEEK)   29
polyethylene (HDPE)   2, 6, 13, 16, 45, 81
polyethylene (LDPE)   2, 6, 13, 16
polyethyleneterephthalate   29
polymer skin   95
polymethylmethacrylate   2
polyphenylene oxide   2, 6, 13, 16, 29
polyphenylenesulphide   6
polypropylene (PP)   2, 3, 6, 13, 16, 26, 27, 28, 29, 45, 67
polystyrene (PS)   2, 6, 13, 16, 18, 19, 20, 21, 45, 67
polystyrene, impact (HIPS)   2, 6, 13
polysulphones   2, 6, 29
polytetrafluoroethylene (PTFE)   45
polyurethane, rigid   117
polyvinyl chloride (PVC)   2, 6, 9, 13, 16, 84, 124
post-mould shrinkage   3
power index/exponent   27, 28
profile plug   91, 92
prototype moulds
  construction   118–120
  materials   117–119
pseudoplastic flow   26
puller pins   65
push rods   47

quality control   115

Rabinovitch correction   30
radiusing   58
register ring   43, 48, 53, 60, 88, 91, 93, 95
retention wedge   101
return pin   43
return wedge   101
Reynolds number   24
ribs   5, 8
riser plate   95
runners   56–58, 92, 95
  dimensioning   28–30
  plate   95
  stripping   88

screw cushion   36
sealing washer   92
self-sealing plug   111
self-tapping screws   9
serviceability   54–55
shear rate   5, 25, 26, 27, 28, 29
shear stress   25, 26, 27, 28, 29
shear thinning   27, 28
shot   36
side core blocks   100, 101
side core pins   97, 98, 99
sinking   3
slug well   62
space block   91
specification sheets   40, 41, 50, 51
split line   10, 43, 44, 53, 57, 87
sprue   57–58
  backing plate   88
  bush   95, 113, 119
  disc   62
  puller   57, 62, 64, 88
  puller bolt   88
  stripper plate   64, 88
standard parts   47, 48, 53, 80, 104, 106,
  107–113, 126
steel alloying elements
  carbon   122, 124
  chromium steel   71, 123, 124, 125
  cobalt   125
  lead   123
  manganese   123
  molybdenum   123
  nickel   71, 123, 124
  silicon   123
  sulphur   123
  tungsten   123
  vanadium   123, 125
steel classification   123–124
steel corrosion resistance   122, 123, 124
steel machinability   124
steel-stress   125
stripper bush   82
stripper plate   47, 88, 91, 95

styrene–acrylonitrile (SAN)   2, 6, 16, 84
sucker pins   65
support puller   46
surface finish
  mould   122, 124
  moulding   11, 115
Swedish Standardisering Kommission
  (ISI)   123

talc   5, 70
tandem safety circuit   103
taper   see draft
tempered hardness   124
tempering   124
thermal conductivity   16, 17, 18
thermal expansion   2
thermal footprint   67, 68
thermal profile   67, 68
thermal properties
  mould materials   70, 71
  polymers   13, 16, 70
thermally sensitive polymers   87, 96
thermocouples   90, 92, 93, 109, 110, 113
tie bar stretch   39
toggle clamp   37, 38
turbulent flow   24

undercuts   9, 10
unscrewing cores   9

vacuum forming   119
vents, venting   45, 64, 79, 80, 81, 91, 115
viscosity of polymer melts   21, 22–24, 25,
  26, 27, 28, 29, 30

wall section (mouldings)   5–7, 69, 114
warping   4–5
water absorption data   2
water, coolant   19, 20, 21, 25
wear of mould parts   46, 47, 48, 55, 65,
  78, 79, 80, 94, 95, 102, 121, 122
weld lines   9, 115
worming   65